SCIENCE
A Many-Splendored Thing

SCIENCE

A Many-Splendored Thing

Igor Novak

Charles Sturt University, Orange NSW 2800, Australia

NEW JERSEY • LONDON • SINGAPORE • BEIJING • SHANGHAI • HONG KONG • TAIPEI • CHENNAI

Published by

World Scientific Publishing Co. Pte. Ltd.
5 Toh Tuck Link, Singapore 596224
USA office: 27 Warren Street, Suite 401-402, Hackensack, NJ 07601
UK office: 57 Shelton Street, Covent Garden, London WC2H 9HE

British Library Cataloguing-in-Publication Data
A catalogue record for this book is available from the British Library.

SCIENCE
A Many-Splendored Thing

ISBN-13 978-981-4304-74-0 (pbk)
ISBN-10 981-4304-74-3 (pbk)

Printed by FuIsland Offset Printing (S) Pte Ltd. Singapore

Dedication

To my former students at the National University of Singapore (NUS) and to my wife Tessa.

I thank my students for their enthusiastic participation in the module "Scientific Methodology" which I taught at NUS for several years. They convinced me that teaching of the subject on the nature of Science was worthwhile pursuing and developing. My wife's continuous emotional support while I was writing this book was (as usual) very valuable and generous.

Preface

"The eternal silence of these infinite spaces terrifies me."
B. Pascal

Science is essentially Man's attempt at understanding Nature and the World around him driven (at least in part) by necessity. Nature is a vast system; it is vast in terms of the number of its constituent parts, of the interactions taking place between the parts and of possible subsystems into which natural world can be organized. This vastness and complexity can engender the feeling of fear (which Pascal experienced when observing the night sky) and instinctive shunning of questions which Man's relationship with Nature fosters.

The main purpose of this book is to acquaint non-scientists (and perhaps some scientists as well?) with Science as a wide-ranging human activity ('a many-splendored thing'). In the 1955 film "Love is a Many-splendored Thing", the director H. King explored problems which cultural barriers driven by irrational adherence to tradition, pose for human emotional relationships. And yet these many-facetted relationships are the source of endless variety and enrichment for human existence. A somewhat analogous situation exists pertaining to modern Science. Advances in Science and Technology have terrified or alienated many people who are afraid to explore the richness and diversity of Nature and of us humans. I hope that this book can allay some of these fears. The word 'Science' used in this book refers mostly to Natural Sciences (mathematics, physics, chemistry, biology) and their derivatives (e.g. biomedical sciences, material science). I hope to present Science in the wider context of Man's perception of the World around him. The book does not expound in depth the Philosophy of Science, Sociology of Science or the History of Science which are established disciplines in their own rights. Instead, the book presents different strands and aspects of Science together in a single volume and demonstrates how overall scientific activity proceeds in practice. For precise definitions and more in-depth analysis of specific topics, references for further reading are provided since it is clearly impossible to cover such a vast subject in a single slim volume. It is therefore advisable to read this book in conjunction with the references provided in order to gain better insight into Science.

The choice of topics and especially of references is selective and not exhaustive. The selection was guided by the current concerns of humanity rather than by academic rigor. It is often considered self-evident that citizens of modern states should posses a modicum of knowledge about literature, art, language, history, law, politics, business, sport and even fashion. Surprisingly and worryingly the same attitude does not extend to scientific literacy. The general public and school leavers are often not only very deficient in basic scientific knowledge and largely unenthusiastic about Science (caused perhaps by the watered down school curricula), but are also ignorant about the basis and limitations of scientific practice. Is it to be wondered then that in the so called Age of Science and Technology various unfounded beliefs and myths come into prominence (mysticism, astrology)? Modern Science probably has, is and will change people's lives to perhaps a greater extent than any other human activity or belief system. It is therefore absurd and dangerous to maintain the current low level of average scientific literacy/awareness of general public. This is a matter-of-fact comment, not an ethical recommendation or an attempt at the glorification of Science. Science is not just a career, hobby or a profession; it is a clear window on the World and on our own humanity. Science is a window not an all round view! It is out of this deep conviction about Science as the outlook on the World that this book has originated.

Igor Novak

Contents

List of Tables

List of Figures

Chapter 1
Introduction

Is Science important?

(all quotations are from [1])

Advanced modern civilization is built on the success of Science and Technology (ST) in harnessing Nature and human endeavors for it purposes. Without it, we'd probably still be a peasant agrarian society. While this isn't automatically a good thing overall, most people would certainly agree that they enjoy improved standards of living which Science has brought us; in fact, the Global South (developing countries) is also moving towards higher levels of science and technology so that they, too, can share such living standards. But is this necessarily the better or the only way forward for humanity? Some authors [2] argue that irrespective of our efforts to reduce or reverse environmental degradation we need to challenge the fundamental dogma of modern civilization. This dogma postulates that unlimited economic growth and increasing material prosperity can and must continue without check, because they are the foundation of the only acceptable lifestyle (or more likely the only one we can imagine at present). Political and business leaders consider ST discoveries to be necessary for global economic development and progress. In turn, the leaders consider such progress to be the only, indispensable guarantor of global political and social stability. Progress is also the guarantor of leaders' own survival as leaders! Whether the human society as an open dynamical system (i.e. system where there is a constant exchange of matter and energy with the physical environment) can ever achieve such stability remains to be seen. We shall discuss this issue in the following chapters. In these authors' view [2] the efforts of environmentalists conform to this dogma rather than challenge it. According to them, the use of Science and Technology to help us get out of our environmental predicament only serves to propagate the ideal of perpetual growth which regards the natural world primarily as a resource for our consumption. According to these views the present crisis (including the 2008 economic crisis) is thus not primarily ecological or economical but the crisis of ethics induced by the confusion of social goals in modern societies resulting from unchallenged orthodoxies (economical and political).

Others would argue that Science answers the 'how', 'what' and 'where' questions, but not necessarily the 'why'. Irrespective of which view of the future development of human civilization we adopt, it is undeniable that ST has immense importance and influence on modern civilization.

Yet for a human activity that has brought us so much, the general population has surprisingly little understanding of core scientific facts. This can be seen from Dr Rita Colwell's presentation on the USA public's understanding of key scientific facts [3]:

- 30.1% believe that genetic engineering will 'make things worse, while 52.8% believe it will lead to improvement, 5% believe it will have no effect and 12.8% simply don't know.
- Only 13% could correctly define what a molecule is, and only 46% knew that the Earth goes around the Sun annually. Even more surprising, only 51% knew that humans didn't live at the same time as the dinosaurs.
- Around 60% believe that astrology isn't at all scientific, while 30% think it is 'sort of scientific', and the rest believe it is either very scientific or don't know.

To make matters worse, many scientists do not have strong grasp of how scientific results can or should be integrated into Society [4] beyond the skill of being able to obtain necessary research funding. This state of affairs may not be scientists' fault, because their efforts have to be focused on the complexities of their work and on the need to align their research with Society's expectations. Nevertheless, this is a concern which may lead to increasing gap between Science and Society and to the impairment of the role which Science needs to play in Society. This role does not involve power politics or an exercise in social influence on behalf of Science. It represents a contribution to solving problems which our increasingly complex and sophisticated civilization generates.

Science and Technology (ST) are important **not only** for scientists and engineers. People encounter scientific information and methods in most careers – consider, for example, various kinds of scientific information that business managers are required to deal with. It is important then for the general public to be able to analyze critically the scientific results that are presented to them.

Equally, we cannot always assume that graduating with a science-related degree will lead to a science-related job. Consider the following graduate destinations of physics students in the UK [5] (from the Higher Educational Statistical Agency's report, *Destinations of Leavers from Higher Education 2005/6*). About 25% remain at University to complete a Masters or PhD when they graduate, but of the rest:

- 20% work in business services,
- 13% work in financial services,
- 11% work in education,
- 10% work in manufacturing, and
- 8%, 7% and 6% work in retail, government and IT respectively.

Similar results have been found in Australia, see URL below. (http://science.uniserve.edu.au/pubs/procs/2008/199.pdf).

The interweaving of ST into every aspect of modern life necessitates the introduction of science subjects for all and not just for those who will find Science professionally useful. However, as Devlin has pointed out [6] the science subjects for non-professionals should encourage science awareness by demonstrating what Science is and how it works.

Why is Science important?

Here are some of the key arguments demonstrating the importance of ST.

- **ST helps us to understand the world**: Scientific literacy is a useful corrective agent. It is the sceptical antidote to mysticism and general confusion between facts and fiction. Such confusion is often created today by extraordinary/ exaggerated claims made in market-driven media. Consider the following examples of unproven processes/events: extraterrestrial intelligence, flying saucers, ghosts, pseudoscientific creationism and so on.

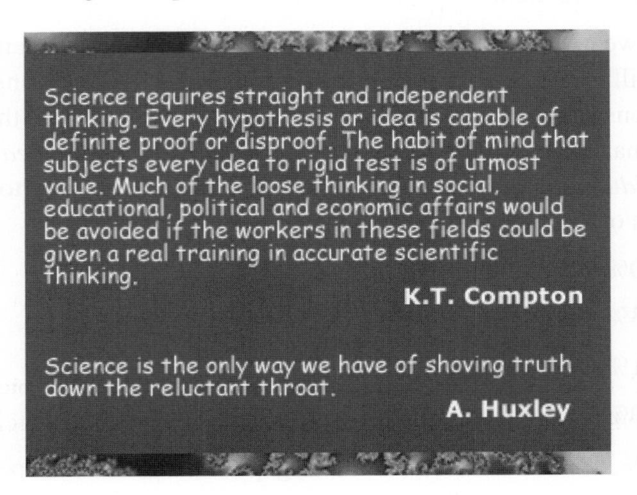

Science requires straight and independent thinking. Every hypothesis or idea is capable of definite proof or disproof. The habit of mind that subjects every idea to rigid test is of utmost value. Much of the loose thinking in social, educational, political and economic affairs would be avoided if the workers in these fields could be given a real training in accurate scientific thinking.
K.T. Compton
Science is the only way we have of shoving truth down the reluctant throat.
A. Huxley

- **ST brings many social changes and challenges**: Citizens can understand and influence these changes only if they are **scientifically literate**, by lobbying for issues regarding environmental policy, social welfare programs, defence spending and consumer choices. ST also has great impact on the physical environment via medicines, transport, communications, IT and so on.

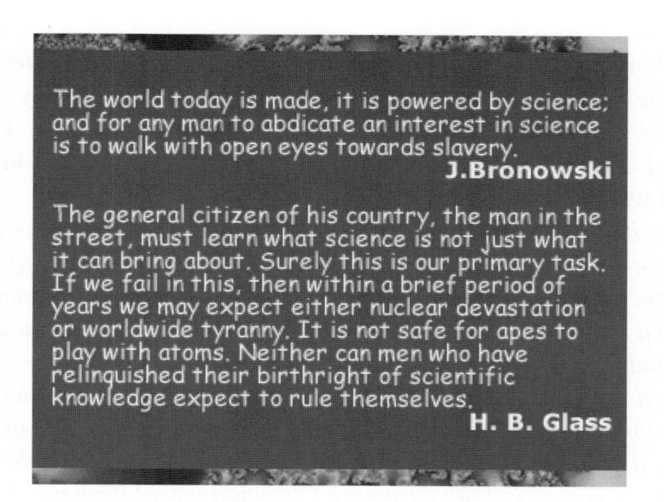

The world today is made, it is powered by science; and for any man to abdicate an interest in science is to walk with open eyes towards slavery.
J.Bronowski
The general citizen of his country, the man in the street, must learn what science is not just what it can bring about. Surely this is our primary task. If we fail in this, then within a brief period of years we may expect either nuclear devastation or worldwide tyranny. It is not safe for apes to play with atoms. Neither can men who have relinquished their birthright of scientific knowledge expect to rule themselves.
H. B. Glass

It is therefore obvious that because ST has such broad *economic and social impact* on modern societies, we should learn about it.

- **ST impacts directly on individual lifestyle and wellbeing:** This happens through discoveries such as genetic engineering, birth control, drugs and cosmetics, diet, GM food, nanotechnology and many others.

- **ST raises deep questions about the direction in which human civilization and individuals' lives should evolve**: Science has encroached into the realms of philosophy and religion. It is involved in answering crucial questions such as:
    - Are human beings special when compared to other biological species?
    - If humans are special what is the origin of that status?
    - How should the relationships between humans be governed?

If we build individual/social ethics solely on biological principles (e.g. evolutionary mechanisms, natural selection) and then proceed by rational deduction from this basis, problems may arise. For example, consider the following quote from Peter Singer, an Australian philosopher:

> *"We should put aside feelings based on the small, helpless and -sometimes- cute appearance of human infants. Laboratory rats are 'innocent' in exactly the same sense as human infants...killing a disabled infant is not morally equivalent to killing a person. Very often it is not wrong at all.*
> *If we go back to the origins of Western civilization, to Greek or Roman times, we find that membership of* Homo sapiens *was not sufficient to guarantee that one's life would be protected...Greeks and Romans killed deformed or weak infants by exposing them to the elements on a hilltop. Plato and Aristotle thought that the state should enforce the killing of deformed infants...The change in Western attitudes to infanticide since Roman times is, like the doctrine of the sanctity of human life of which it is part, a product of Christianity. Perhaps it is now possible to think about these issues without assuming Christian moral framework that has for so long, prevented any fundamental reassessment."*

The problem is not intrinsic to scientific theories themselves i.e. to Theory of Evolution which correctly describes the variety of species and their origin. It is the derivation/extrapolation of metaphysical or ethical ideas from scientific theories which may be unjustified!

- **ST has influenced Art.** There are many examples of this influence. For example, new scientifically based technologies in sound production have spawned new forms of musical art with new tonalities. In film industry and literature there is a development of computer animation and the flourishing of science fiction novels. Finally, the scientific images from the microscopic and macroscopic worlds are unusual and artistically highly stimulating. Ball [7] has suggested that many unusual images generated thorough scientific research can be used as a source of inspiration for visual theatre. Images like those given in Figure 1 are of objects invisible to naked human eye. They were taken from the mathematical world (Mandelbrot set), the world of very large (galaxy) and of very small, nano-world ("quantum corral") and certainly exhibit artistic potential. Visual imagery or sound spectra are not the only examples of how Science can influence Art. Some modern plays have Science as one of their main themes [8]. They explore how in a human individual the psychological and emotional perceptions of the World intertwine with perceptions acquired from Science.

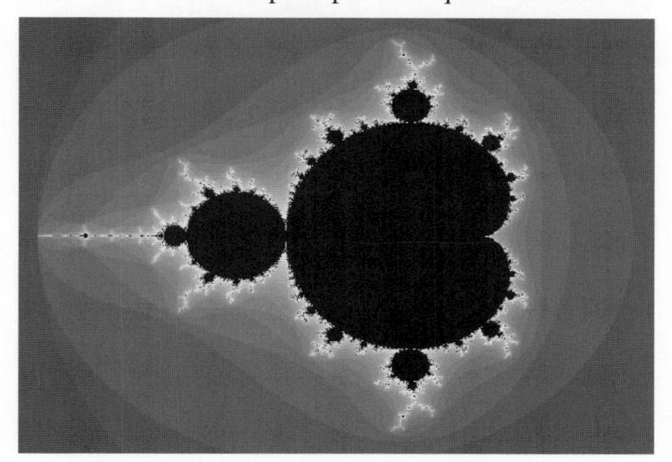

source: http://conspiracyfactory.blogspot.com/2008_04_01_archive.html (Mandelbrot set)

source: http://www.freeimages.co.uk/ (Galaxy)

source: http://employees.csbsju.edu (Quantum corral)

Fig. 1. Scientific images with artistic appeal.

Reflection:

Consider the often encountered comment: "ST is very important because it has given us an almost unlimited mastery of Nature." Is this notion true at the most fundamental level?

Tip: Consider the relationship between Humanity and Nature. Are we part of Nature?

What is Science?

In order to understand how Science can be misrepresented and misused, we must first understand Science itself. The word Science originates from: *scientia* (Lat.) = knowledge.

Science is a body of knowledge (facts, theories) **AND** a set of methods used to acquire knowledge/understanding. In short, Science is a way of thinking and a world view!

Until the 1830s, the word Science was applied to any organized body of knowledge. The activity we consider as Science today used to be carried out by individuals who called themselves 'natural philosophers'. In contrast, Art implied creative technique and human feelings. Today, the word Science encompasses Chemistry, Physics, Biology, Social Sciences as well as other new disciplines like environmental science, material science etc. Let us consider some aspects of modern Science:

- **Science comprises search for order and rational explanations using observations and logic:** The main purpose of Science is to trace, within the apparent chaos and flux of phenomena, a consistent structure with order and meaning. This effort conforms to the philosophy of rationalism (rational means conforming to or understandable by human mental faculties). The purpose of scientific understanding is to coordinate our experiences and bring them into a logical system or framework. The search for regularity and order *necessarily* involves certain amount of simplification/reduction of the information received through our senses or instruments. We can use mathematical language to describe scientific activity as: 'many-to-one mapping' i.e. as mapping between the elements of the domain (human environment) and range (human cognitive system) [9]. Any such mapping *necessarily* involves data compression and reduction. We shall mention maps in the context of scientific knowledge again in this Chapter and the next. This reduction exercise of course is one of the reasons which preclude any aspirations which Science may have towards attaining the 'absolute Truth'.
- **Science is driven by our need and desire to control the (unknown) environment:** Throughout history, intellectual and scientific efforts have been directed towards the discovery of patterns, systems and structures, with special emphasis on order. Order allows human individuals and societies to adapt to their environments and to ensure their own physical existence and reproduction.

- **Discovery of patterns and structures is a key to scientific problem solving:** The problem solving leads to understanding and control of the environment in which humans live. Those who pursue such activity are known as scientists. The main occupation of a scientist is problem solving with the goal of understanding Nature and Universe and discovering patterns and regularities in the known World.

- **Science is based on the idea that Universe is ordered and can be comprehended using Reason; it is subject to rational explanation:** Science is founded on the belief that the World is rational in all its observable aspects. However, unlike some other beliefs, this belief system can be verified by observation of and reference to many natural phenomena. It is a belief nonetheless, because we cannot observe or comprehend all natural events. This opens the door to the possibility that there may be some facets of reality which lie beyond the power of human reasoning, that there are things whose explanations we could not grasp or phenomena for which we have no explanation at all. However, such facets are not the subject for scientific inquiry. The fact that the World is rational is connected to the fact that it is ordered. An ordered World is a prerequisite for the existence of Science. We shall discuss the meaning of explanation in Chapter 2, but suffice it to say that every 'explanation' always involves some form of interaction between human brain and the physical World.

- **In Science, there is an emphasis on data/evidence and prediction, not on absolute, metaphysical answers:** The keystone to Science is proof or evidence/data, which is not to be confused with certainty. Except in pure mathematics, nothing is known for certain, although much of our knowledge is certainly false. Central to the scientific method is a system of logic.

- **Science is not about Truth, it is about testability and internal consistency:** Science is a dialogue between mankind and Nature. It is not a perfect instrument for acquisition of knowledge, but it

> provides something that other methods fail to get: practical results. Science is a 'candle in the dark' which often dispels irrational beliefs or superstitions. Science does not, by itself, advocate courses of human action, but it can certainly illuminate possible consequences of certain actions. In this regard, Science is both imaginative and disciplined, which is central to its powers of prediction. Science differs from pseudo-science by the principle of falsifiability (the notion that ideas must be capable of being proven false in order to be scientifically acceptable; see Chapter 2).

How does Science differ from other fields of human activity and culture e.g. Art? Both activities use imagination and creativity, but outcomes of artistic activity are mostly free and unconstrained. Ideas in Science must conform to experiment and observation. In other words, scientific outcomes can be tested and must conform to reality as expressed through experiment and observation.

Science is a system of knowledge that is concerned with the physical world and its phenomena and entails unbiased observations and/or systematic experimentation. In general, Science involves a pursuit of knowledge covering general truths or the operation of fundamental laws of Nature. Flaubert has summarized the answer to: "What is Science?" very well (see the citation below).

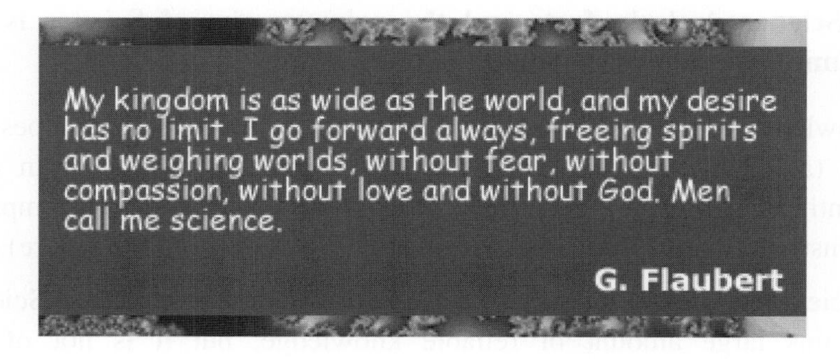

My kingdom is as wide as the world, and my desire has no limit. I go forward always, freeing spirits and weighing worlds, without fear, without compassion, without love and without God. Men call me science.
G. Flaubert

Misconceptions about Science [10,11]

Despite the exceptionally important role which Science plays in modern civilization general population perceives Science as being related to the following three entities:

- a set of facts and theories that explain facts
- a particular approach/method to problem solving
- a set of tasks performed at scientific institutions

This narrow view of Science leads to some common misconceptions which we shall discuss next. Understanding these misconceptions is very important for several reasons. Knowing what Science can or cannot do is necessary if we are to use and support Science effectively. This understanding is also important to avoid various delusions and errors which may originate from misreading the scientific message. In modern World mass media are often the main source of information about Science. The media are market driven and select the information which is presented to the public according to such market criteria rather than according to intrinsic accuracy of the information or its scientific value (media wants to 'tell a good story'). We shall discuss specific misconceptions (A-K) next.

A) Science deals in facts and the primary aim of Science is the accumulation of experimental facts and observations. *False.*

Knowledge is much more than a collection of facts. Science does not deal (as F. Bacon assumed) with 'bare, unchangeable facts'. In fact, scientific knowledge is more map-like than fact-like; it can be compared for instance to the public transport map in Figure 2 (from Singapore).

If scientific knowledge is not fact-like, is it then unreliable? Science contains large amount of reliable knowledge, but it is not of the philosophical and metaphysical type, e.g. 'mass exists'. Rather it is of the type, 'If certain body has the property (called mass) and moves towards the wall, it will show an impact, or dent on the wall which will be proportional to its mass'.

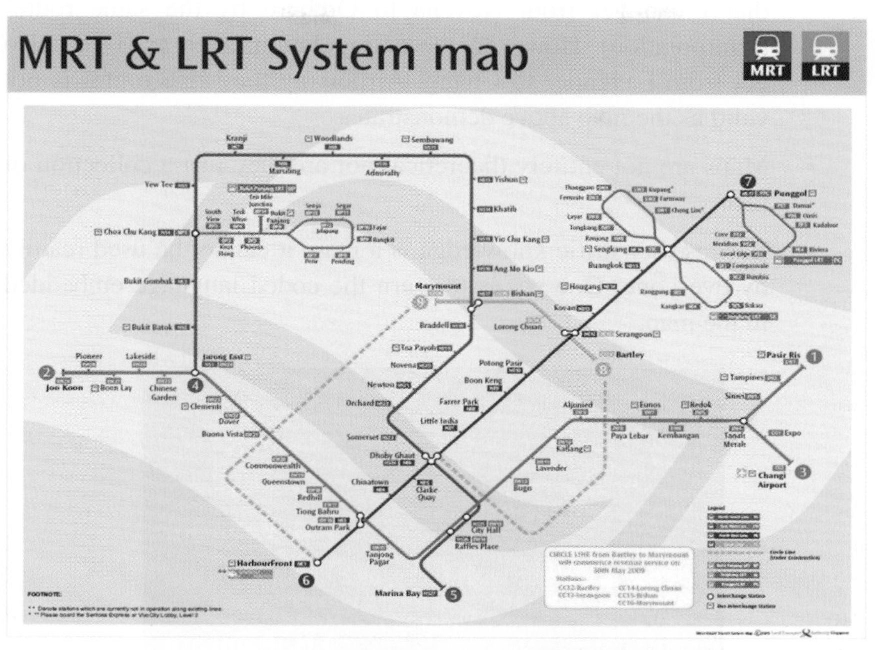

source: http://www.transitlink.com.sg/images/eguide/mrt_sys_map.htm
Reproduced with permission from the Land Transport Authority, Singapore

Fig. 2.

In what ways is Science like a map?

- A map is not Reality itself, but a reflection of and guide to Reality. Map doesn't give all the details of Reality, only the crucial ones. In that way, maps are very good at providing some kinds of information (e.g. how to get from Kranji to Novena station), but useless at providing other kinds (e.g. what buildings shall I find at Novena?).
- Maps are subject to change and so is Science. Places get renamed, stations change, and so on.

- Maps contain a lot of information. For example, 'If I can get from Kranji to Somerset by a certain route, then I can assume that I can get from Novena to Orchard by the same route' (interpolation). However, the extrapolation, such as 'I can also get from Lavender to Changi Airport by the same route' is not valid as the map above demonstrates.
- Maps are not entirely theoretical nor are they just a collection of facts.
- Because scientific knowledge is a map, it cannot be used readily by everyone. One needs to learn the coded language embedded in the map.

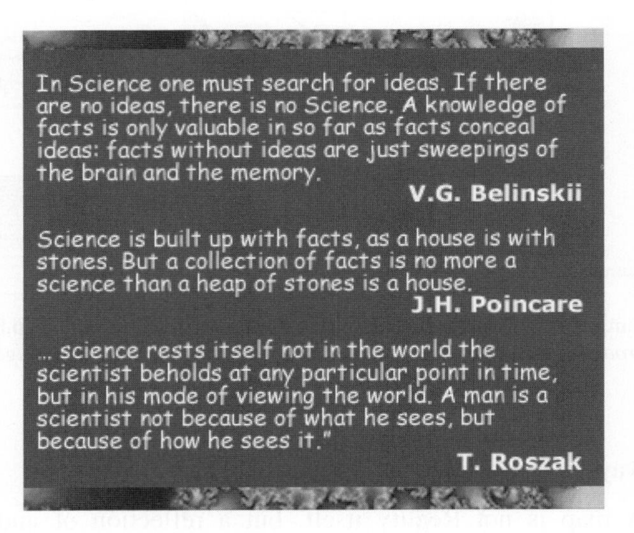

In Science one must search for ideas. If there are no ideas, there is no Science. A knowledge of facts is only valuable in so far as facts conceal ideas: facts without ideas are just sweepings of the brain and the memory.
V.G. Belinskii
Science is built up with facts, as a house is with stones. But a collection of facts is no more a science than a heap of stones is a house.
J.H. Poincare
... science rests itself not in the world the scientist beholds at any particular point in time, but in his mode of viewing the world. A man is a scientist not because of what he sees, but because of how he sees it."
T. Roszak

B) Science is a product of Western cultural tradition and thus has little relevance for other cultures. *False.*

Science deals with important aspects of Reality which can be accepted and verified universally by members of any culture, religion or social group. The same is not true of other disciplines such as Ethics, Law, Art and Religion which deal with various aspects of human character and human psychology.

C) Scientific knowledge is the ultimate attainable 'Truth'. *False.*

Scientific laws and discoveries are subject to modifications; there are **no absolute truths** in Science. This misconception is related to 'scientism'. Nature and Life are not 'clockwork mechanisms' or finished products and therefore cannot embody the 'final Truth'. They are processes involving perpetual transformations of matter and energy. This ensures progress/evolution and that the future will be different from the past. Existing scientific knowledge may be partly wrong or incomplete. Individuals who perform Science (scientists) cannot be entirely objective in their work and for this reason all human activities and institutions (including the scientific community) are prone to errors. The citations below express another important problem with the notion of Science as a path to absolute 'Truth' (scientism).

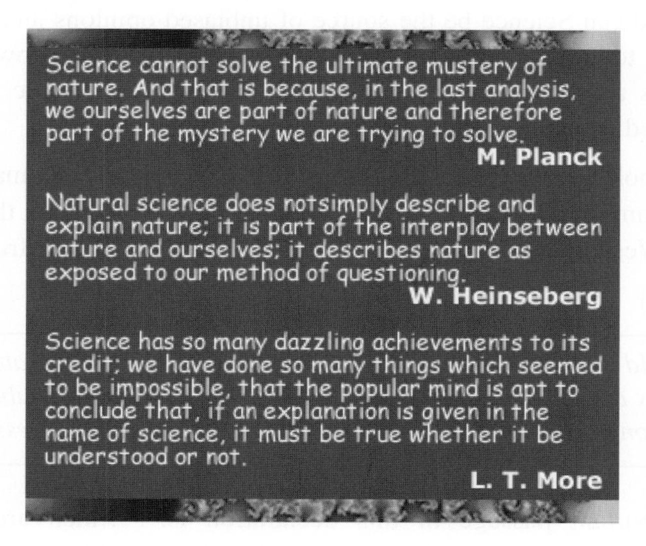

Science cannot solve the ultimate mustery of nature. And that is because, in the last analysis, we ourselves are part of nature and therefore part of the mystery we are trying to solve.
M. Planck
Natural science does notsimply describe and explain nature; it is part of the interplay between nature and ourselves; it describes nature as exposed to our method of questioning.
W. Heinseberg
Science has so many dazzling achievements to its credit; we have done so many things which seemed to be impossible, that the popular mind is apt to conclude that, if an explanation is given in the name of science, it must be true whether it be understood or not.
L. T. More

D) Science should be primarily concerned with solving practical problems; Science as a 'toy factory'. *False.*

Science seeks understanding. Technological advancement does not necessarily lead to better understanding of Nature or to the improvement of human conditions. Technology is the necessary link between Science and Society. However, Science can and should help us to understand ourselves and Nature and avoid errors in our judgments.

Science is today often constrained by economic and social considerations and by political correctness. This is a consequence of the prevailing political dogma of unlimited economic growth and material prosperity and Science as the main tool for achieving this aim. This dogma diminishes the potential of Science as an independent source of knowledge and at the same time weakens human relationships and the fabric of Society at large. The evidence for the claim about material prosperity and weakened social fabric comes from recent research work [12]. Why is it important not to constrain Science by the currently prevailing political and economic dogmas? Because only when unfettered can Science be the source of unbiased opinions and effective solutions to current problems. The case of spread of 'mad cow' disease in UK is a good example of what happens when scientific advice is obfuscated for political or economic gains.

Let us look again at the relationship between political dogma and the constraining of ST. One of the famous political documents in the history of the Western world (USA Declaration of Independence from 1776) states:

> *"We hold these truths to be self-evident, that all men are created equal, that they are endowed by their Creator with certain unalienable rights, that among these are Life, Liberty and the pursuit of Happiness....."*

The important passage in this declaration is: "...men are created equal..". Science in fact demonstrates that this "truth" is neither self-evident nor is it based on reality. Human beings are not created equal, they have different abilities, personalities etc. This Declaration represents human aspirations, hopes and a form of ethics rather than physical reality. The spirit expressed in this Declaration has contributed to the rapid development of modern liberal democracy, market capitalism and consumer society which has successfully used ST for its own ends. The problem is that this political ideal, expressed as personal independence and individualism and embedded in Western collective consciousness is often a taboo carried to the extremes and thus hinders understanding of human interdependence and the need for cooperation on the global scale. Science tells us that ***men are not born equal, but are born dependent on each other*** i.e. on other fellow human beings for our collective and individual survival. This is consistent with our biological nature as social primates. The correctness of this scientific view has been underlined by modern globalization processes and the appearance of global problems (climate change, terrorism, economic disparities, competition for natural resources, 2008 financial crisis). This is an example where proponents of the prevailing political and economic dogmas use Science as a tool for preservation and continuation of these dogmas. How does this happen? ST is used to stimulate unchecked exploitation of natural resources rather than to devise solutions for sustainable development and self-understanding of our civilization. There have recently been increased efforts to use ST as a 'cure' for problems which our own civilization has created. Once again, while this is certainly an important task for ST it is not Science's primary task. If Science was left to operate with fewer economic and political constraints perhaps its findings would have helped us to avoid some of the current global problems in the first place.

E) Science has a strong influence on Society and is not politically motivated. *False.*

The influence of Science is smaller than is widely believed (see the immediately preceding discussion), while the influence of Technology is much greater. Society in fact has a strong influence over Science! For example, consider the following extract from the article published in *Chemical & Engineering News* (November 24, 2003, pp. 25) as an example of political interference in Science:

ACS imposed moratorium on some foreign papers (since rescinded)
The American Chemical Society (ACS) has imposed a moratorium on publishing journal articles that are written by authors in Cuba, Iran, Iraq, Libya, or Sudan. Robert D. Bovenschulte, president of the society's Publications Division, writes in a memorandum to ACS journal editors that "this action, which I hope will be temporary," has been taken "with great reluctance." The moratorium follows a ruling by the Treasury Department, whose ***Office of Foreign Assets Control (OFAC)*** *"advised that papers from these nations may be published but only if no editorial services are provided," Bovenschulte writes. Such services could be deemed to violate U.S. trade sanctions against the countries. The Treasury Department ruling was prompted by actions of the* **Institute of Electrical & Electronics Engineers (IEEE)***, which is attempting to obtain a license from OFAC that will allow IEEE to edit articles by authors in the affected countries. If IEEE, which Bovenschulte notes has "taken the lead with the government on this problem," can obtain such a license, "ACS can then use that precedent to make the same case to OFAC and so obtain our own license."*

F) Science and scientific method alone determine what is true and provide complete and satisfactory explanations of all phenomena. *False.*

This misconception, similar to C is called 'Scientism' or 'Naturalism'. Human knowledge of the natural world is not exhaustive or conclusive. It is not exhaustive, because obviously there is a lot that we do not know about the physical world. It is not conclusive because (as shall be discussed in Chapter 2) World is not a finished object, but a chain of irreversible processes. Because many of these processes are irreversible we cannot always predict their final outcome. Furthermore, human knowledge is filtered through human theories and instruments. Science is a human activity and scientific theories are human constructions. This does not imply that scientific theories are wrong, only that they are limited in scope. Non-scientific, metaphysical concepts are not *necessarily* meaningless or untrue although they may be.

The diagram below summarizes this idea. We often do not experience Nature directly, but through our senses, models and instruments. We do not 'see' that Earth is spherical, but we piece together several observations (e.g. the changing contour of the ship which sails away from us) and make several assumptions (e.g. that light travels in a straight line) before we can draw the conclusion about the Earth's shape.

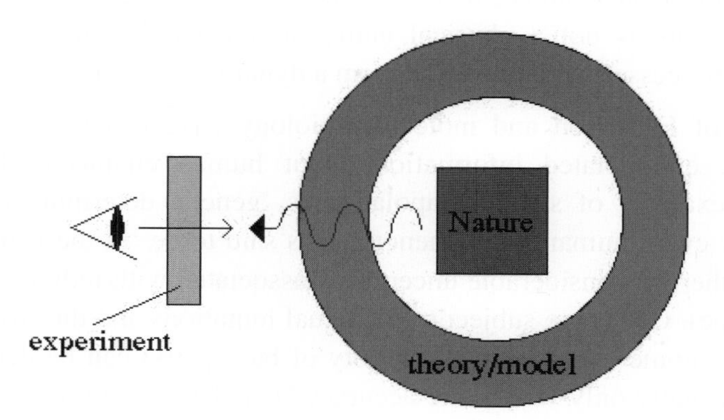

source: http://abyss.uoregon.edu/~js/21st_century_science/lectures/

Truth Paradox about scientific knowledge is that there are scientific claims which are not true and claims which are true, but are not scientific!
Examples:

- phlogiston theory was scientific and yet was not true representation of natural process
- statement: "France won 98 World Cup" is true, but is not scientific
- The claim of 'naturalism' which regards Science as the sole source of Truth has been subject to lengthy discussions. The misconception is related to the old debate about "nature vs. nurture". This misconception is very important because it impacts upon our notion of what human nature and identity are [13] and (via extrapolations from scientific knowledge) gains

ethical implications. Are we humans biologically or culturally determined? In humans, the intellect has subdued emotions and instinct in many (but not all) aspects of behaviour. Does this fact represent a difference of degree or a difference of kind? Is human nature an emergent property which can be understood only within the evolution of complex dynamical systems i.e. individual human beings? It needs to be reiterated that human being is not a physical entity or a body, but rather a set of processes which together form a dynamical system.

Theory of Evolution and molecular biology have often been used to generate extrapolated information about human character. The best known example of such extrapolation is 'genetic determinism' where virtually every human trait (phenotype) is said to be 'in the genes' even though there is considerable uncertainty associated with individual genes (genotype). Genes are subject to perpetual mutations and the structure of human genome (sequence and identity of base pairs) can be determined experimentally only with finite accuracy [14]. The accuracy is related to the product of mutation rate and the genome size. This observation suggests that one-to-one correlation between genotype and phenotype is impossible even in principle. This finding obviously invalidates any form of genetic determinism. This 'genetic uncertainty principle' is thus analogous to Heisenberg uncertainty principle in quantum mechanics where particle's properties are also determinable only up to finite accuracy.

The reason for most extrapolations made from scientific results is due to our human need and desire to know and thus control our environment to our advantage.

G) Science and scientific method can be used to prove or disprove the existence of God! *False.*

If God's relationship with the physical Universe could be understood scientifically that would imply that God himself is part of the same physical Universe. God, according to monotheistic religions, is the author/creator of the Universe and not an entity within the Universe itself or an entity subject to Natural laws. God's relationship with the World and mankind is therefore a philosophical/metaphysical and not a scientific question.

There are two extremist or 'fundamentalist' positions regarding this question:

- creationism (extrapolation from Religion to Science)
- reductionism (extrapolation from Science to Religion)

The two positions are almost mirror images of each other and tend to justify each other's existence and prejudices. The two positions shall be discussed further in Chapter 2.

It is **not** the purpose of religious bodies to interfere with the workings of Science by commenting on the content of specific scientific theories. Such actions were sometimes practiced in the past (e.g. trial of Galileo by the Inquisition) because of the mistaken belief that Science has religious implications (or conversely that Religion has scientific implications). Today similar beliefs are embodied in the Creationist movement.

It is **not** the purpose of Science to devise metaphysical theories and world views based on derivations and extrapolations from scientific theories. Science is not a metaphysical/philosophical world view, but a series of procedures and techniques which are used to obtain reliable knowledge about the physical world we live in.

Religion and Science answer different questions about the World [15]. Whether there is a purpose to the Universe or a purpose for human existence are not questions for Science to answer. No single form of knowledge provides answers to all human questions (see citations below). It is often thought that Evolution is incompatible with the notion of God. However, Haught has proposed that this is not so [15]. Religion is all about (unknown) future. Religion according to this view does not consider the past as the sole determining influence on the present and future events nor is Religion about the absolute, ideal World to which believers must aspire. Religion is about the believers and God participating together in the world creation. Religion is about unfolding the mystery and opportunities of Life and this view can be readily accommodated by Theory of Evolution. Interestingly, God thus becomes a "giant attractor" (to borrow the expression from the theory of dynamical systems) influencing all Life in the Universe, but not in a rigid, predetermined way. While this idea seems to reconcile Evolution and Religion, it remains speculative since there is no conceivable way in which it can be tested.

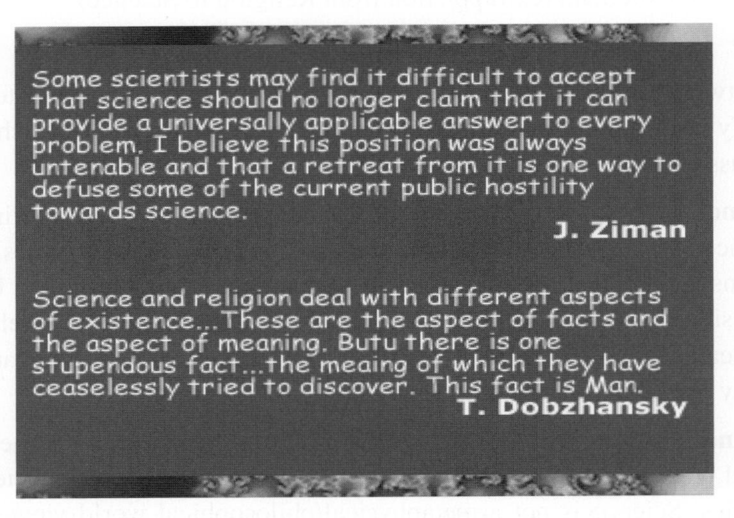

Some scientists may find it difficult to accept that science should no longer claim that it can provide a universally applicable answer to every problem. I believe this position was always untenable and that a retreat from it is one way to defuse some of the current public hostility towards science.
J. Ziman
Science and religion deal with different aspects of existence...These are the aspect of facts and the aspect of meaning. Butu there is one stupendous fact...the meaing of which they have ceaselessly tried to discover. This fact is Man.
T. Dobzhansky

As a final comment, let us point out that if God's existence could be 'proved' or 'disproved' there would be no Religion. Proof or disproof obviates the need for belief and replaces it with certainty. Attempts to use Science in support of or for denial of Religion probably reflect residual doubts about Religion or Atheism in the respective communities of believers and atheists. The believers who may have doubts about their Faith seek additional assurance to support their beliefs and select bits of scientific knowledge/results for that purpose. The atheists are irritated by stubborn, 'incomprehensible' persistence of religious beliefs in Society in spite of atheists' conviction that it should be only too obvious how nonsensical and harmful Religion is. Thus atheists also use Science to try and prove their belief in the non-existence of God, to discredit Religion at the fundamental level and to find assurance for their own materialistic worldview (see Chapter 2 for the discussion of holism and reductionism).

Proposed 'scientific' arguments for God's existence

Natural Theology claims that order in Nature indicates the existence of an intelligent designer who created an ordered, functional system (like a watchmaker who creates a complex, ordered product: watch). The analogy is based on the extrapolation from our own experience where various intricate devices imply the action of intelligent maker. This argument has been discredited by the Theory of Evolution (TOE). TOE proposed a mechanism capable of generating order and complexity in dynamical systems without influence of any external intelligent agent.

Life is not a closed, perfected system like a mechanical clock [15]. It continuously adapts itself to the changing environment. If Life was a designed system, the 'designer' would have to continuously tinker with the system to keep it working. Instead, natural selection processes perform these adjustments automatically. How does this happen? This happens because natural variations ('copying errors') occur during the replication of organisms. This error propagation accompanies every process which duplicates objects. Let us remember that manufacture of any product or any act of measurement can never produce the exact copy of an object or the measurement result. This is due to the ubiquitous presence of random errors! The environment in which these imperfect replicas reside imposes restrictions on their usefulness or survival. Those copies which are not well adapted or useful (in case of industrial products) cannot exist under the given constraints and are eliminated. The environment thus imposes 'direction' into the development of entities and processes. The presence of 'direction' seems to indicate the presence of a Designer or a 'purpose', but there is none. It is important to add that natural selection acts only on populations/groups of objects which are similar, but not identical. Natural selection process cannot act on single entities or on the set of identical entities.

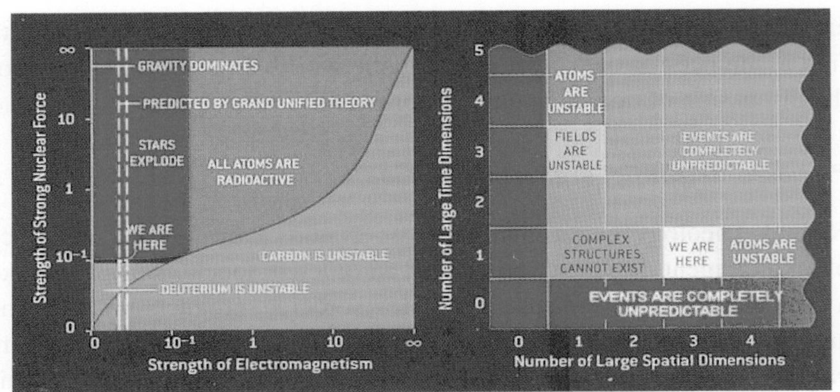

source: http://universe-review.ca/I02-21-multiverse3.jpg

Fig. 3.

Anthropic principle: "Imagine a universe in which one or another of the constants of physics is altered by a few percent one way or the other. Man could never come into being in such a universe. That is the central point of the anthropic principle. According to this principle, a life-giving factor lies at the centre of the whole machinery and design of the world." Barrow and Tipler

The anthropic principle (AP) is a theory relating our existence to initial, physical conditions prevailing at the time when Universe came into being. According to AP, the Universe was 'fine-tuned' to make carbon based Life possible. Any slight variation in these conditions would have made Life and Science impossible and Universe unobservable. These initial conditions are related to values of several fundamental physical constants (see Figure 3):

- **Gravitational constant,** determines the strength of gravitational forces. If its value was lower than it is, stars would have insufficient pressure in their interiors to overcome Coulomb barrier and start thermonuclear fusion (i.e. stars would not shine). If it was higher, stars would burn up too fast and use up nuclear fuel before generating element carbon and giving Life a chance to evolve.

- **Strong and weak nuclear force constants,** determine the magnitude of forces which hold particles together in the atomic nuclei. If they were weaker, multi-proton particles would not hold together, and hydrogen would be the only element in the Universe. If they were stronger, all elements lighter than iron would be rare. Also there would be less radioactive decay, which heats the core of the Earth and facilitates volcanic activity. The heat also supports hydrothermal vents (deep on the ocean beds) where first Life is believed to have started.
- **The electromagnetic coupling constant,** determines the strength of the electromagnetic force which binds electrons to the nucleus and makes the existence of atoms possible. If the constant was smaller, no electrons could be held in orbit around the atomic nucleus and thus no atoms could be formed. If it was larger, the electrons of one atom would not chemically bond to the neighbouring atom to form molecules. Without atoms there would be no molecules and without molecules there would be no Life (as we know it).

However, AP may refer to a coincidence and is **not a proof** of God's existence! Furthermore, some modern research has indicated that these fundamental 'constants' have had different values in the past.

H) There are no limits to what Science can do. *False.*

Limits exist as to what Science and Technology can do or discover. Science exists precisely because there are limits and constraints on what is possible in Nature. Laws and constants embody such constraints. In the totally random world there would be no Science and no Life! (see Chapter 3 for further discussion)

I) Opinions of great scientists are authoritative declarations regarding scientific and general truths. *False.*

Science is not a monolithic organization or activity. Science is a collection of diverse 'sciences', without a universal, rigid method or mechanism for automatic 'generation' of truths. Great scientists (e.g. Nobel laureates) usually have in-depth knowledge in only one narrow field. Scientific discoveries depend not only on intellectual ability, but also on luck, having the right experience at the right time and so on. Therefore, practice of Science requires such intense mental concentration and time investment that scientists cannot easily obtain useful perspectives on wider issues, such as the social role of Science, ethics and happiness. Hence technically successful scientists are not well qualified to answer non-scientific questions.

The public wants definitive, precise answers and recommendations about fundamental questions regarding practical applications of ST or sociopolitical implications of various projects and ideas. Public does not wish to know intricate technical details, but demands instead simple, authoritative answers (even when such cannot be provided). Great scientists are supposed to be great authorities in the field of Science so authoritative answers are *expected* of them and *interpreted* as such.

The source of this misconception is that the public conflates misconceptions **C, D, F** and **I** and endows Science practitioners with supreme authority and monopoly regarding the Truth.

J) All advances in scientific knowledge can be converted to useful applications. ***False.***

This misconception is tacitly perpetrated by scientists who often tend to portray their research in this light in order to secure research funding. The misconception (see the cartoon of money-making machine in Figure 4) drives social and political pressures for Science to be accountable in the same way as social services are to taxpayers or businesses to shareholders. It is impossible to predict which fundamental research areas may lead to useful applications (for example, laser was considered a useless curiosity when it was designed!). This misconception fails to distinguish between 'textbook science' and 'frontier science', (see Chapter 2), it confuses achievements of Science with achievements of Technology and disregards the existence of conflicts of interest. Science is open and knowledge-driven; Technology is secret and profit-driven, hence there can be possible conflicts of interest. The conflict of interests may arise for example, when scientists who designed a new drug for a pharmaceutical company are also involved in testing its side-effects or when scientists in military industry assess the effectiveness of their newly designed weapons.

source: http://www.flippingmind.com

Fig. 4.

This misconception is relevant to:

- corporate ventures between university and industry which follow the simplistic reasoning and expectations: scientific advance → new technology → profit.
- goal-oriented research towards narrow, specific targets e.g. cancer cure. Such research requires as a prerequisite expansion of basic scientific knowledge which must show how something can be done. For example, a much more comprehensive knowledge of the cell's functioning must be attained before a

cure is likely to be found. Goal oriented research which lacks fundamental understanding of relevant processes has dubious prospects unless serendipity intervenes.

The aforementioned comments do not suggest that scientific knowledge should not be converted to useful applications. The comments only suggest that this conversion is not always feasible, may not be feasible immediately or may involve undesirable effects.

K) Increased scientific literacy will enable people to make better decisions, will supplant superstition and make people behave more rationally and ethically. ***False.***

Scientific literacy (SL) is important, but not because it is a tool for any specific goal [16]. A scientifically literate population is likely to do many things better and live happier lives, but it cannot be automatically predicted which things will be done better and to what extent. For example, knowledge about obesity and smoking does not automatically lead to healthier diet or to abstention from smoking. Despite scientific education, general public's view is often confused and involves irrational choices.

> *"I believe in a bit of Scientology, Catholicism, Judaism and the Eastern philosophies. I take a bit of each, I am a hybrid."*
>
> *Nicole Kidman*

The quote from actress Kidman shows that confusion and irrationality are very present even in the scientific age. She is obviously hedging her bets!

SL should therefore comprise some understanding of basic Science, but **also** the understanding of what scientific activity is and what is the place of Science in modern society and culture [6]. Science should not be equated with absolute Truth, nor with other desirables like peace and happiness. Science is not a gateway to Paradise nor is it a guarantor of better, more stable Society (as the citations below indicate)!

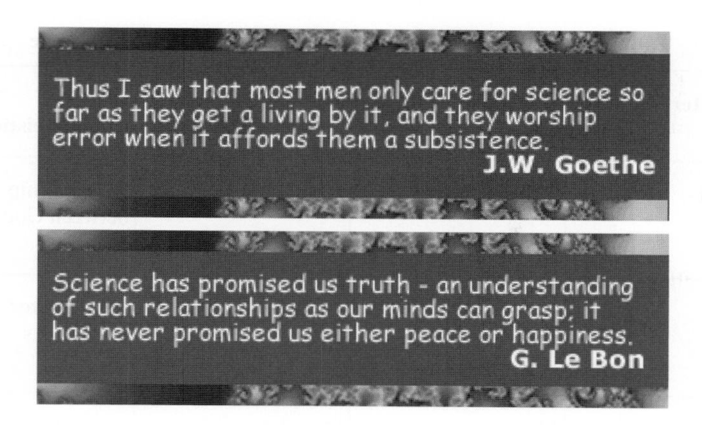

Thus I saw that most men only care for science so far as they get a living by it, and they worship error when it affords them a subsistence.
J.W. Goethe
Science has promised us truth - an understanding of such relationships as our minds can grasp; it has never promised us either peace or happiness.
G. Le Bon

Science and Culture

CULTURE is a shared and organized pool of various forms of knowledge and experience.

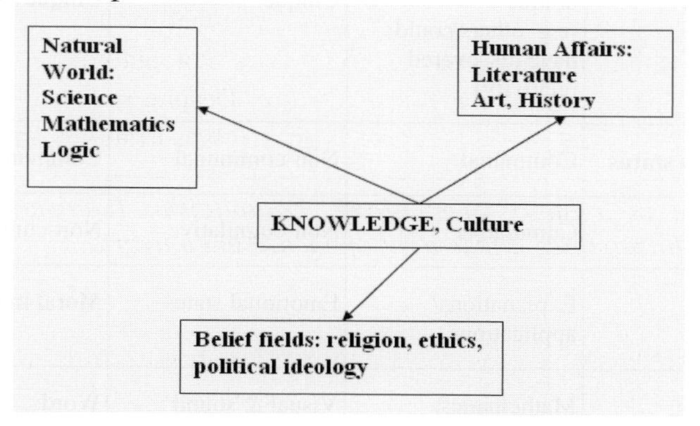

The diagram above is an example of a *concept diagram* or *concept map*. The arrows show how the entities are related – how they depend and influence each other. This concept diagram shows that Science is an element of Culture. Science differs from other cultural elements (human affairs and belief fields) because it incorporates 'reality therapy'. This means that scientists reject ideas and concepts which do not work in practice or are inconsistent with empirical results.

The following Table compares three important elements of Culture.

Table 1: Comparison of Science with other elements of Culture

Characteristic	**Science**	**Art**	**Religion (monotheistic)**
Subject	Natural phenomena	Mankind	Relationship between God and Mankind
Source	Observations experiments logical reasoning	Human emotions	Revelations/ holy books
Methods	Measurement, logical analysis, Theory construction	Stimulation of senses	Textual interpretations, ritual, reflection
Creator-content relationship	Impersonal/not unique (e.g. others could have discovered relativity)	Personal/ Unique	Personal/ Unique
Community status	Communal	Non-communal	Communal
Progress	Cumulative	Non-cumulative	Non-cumulative
Results	Explanations/ applications	Emotional states	Moral imperatives
Expression	Mathematics, symbols, logic, metaphors	Visual & sound stimuli,metaphors	Words, metaphors
Limitations	No set of goals/values, no metaphysical questions	Anthropocentric, socially constructed	Mechanisms unexplained
Examples of Distortion	eugenics	pornography	'holy war'

The quotations below also illustrate relationships between these cultural elements.

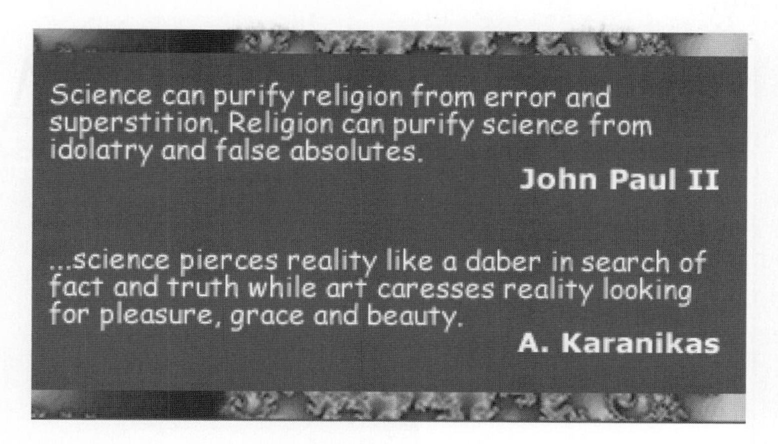

Science can purify religion from error and superstition. Religion can purify science from idolatry and false absolutes.
John Paul II
...science pierces reality like a daber in search of fact and truth while art caresses reality looking for pleasure, grace and beauty.
A. Karanikas

Possible distortions and abuses of a particular cultural element do not invalidate that element or make it undesirable as a whole. For example, the fact that some paintings can be considered pornographic by some viewers does not imply that we should reject all Art! The fact that Religion may be abused for political agendas by some ('holy war') does not imply that Religion is socially harmful. Science itself has been abused in the past to justify 'racial cleansing' (selective breeding of healthy humans, killing of retarded persons etc.) in the form of eugenics. This does not imply that we should reject Science!

What is the principal difference between Science and other human activities?

Science uses reproducible, controlled experiments (as part of the 'scientific method') and draws conclusions on the basis of experimental results **and** logical reasoning **alone**. This is different from other human activities shown in the Table above. Humans often follow emotions and moral principles when performing actions and making decisions about their lives.

General knowledge versus scientific knowledge

Sources: http://www.time.com/time/covers/0,16641,07-01-1945,00.html
Reprinted with permission from J.Phys.Chem.A November 19, 2009, 113(46) Copyright 2009 American Chemical Society

Fig. 5.

Vivid comparison between these two types of knowledge may be gleaned from the cover pages and contents of publications reporting information corresponding to these two types of knowledge (e.g. publications: "Journal of Physical Chemistry" and "Time" magazine in Figure 5).

In scientific journals (e.g. "Journal of Physical Chemistry") the published information is organized according to the following principles:

- topics and problems reported in scientific journals are only those which are amenable to scientific analysis (i.e. which are quantifiable, reproducible). The topics in general magazines do not fall into this category. For example the public figures and issues highlighted are often related to politics, fashion, sport and show business. For example, Anita Colby ('Time' magazine) was a model and actress.

- aesthetic preferences (female beauty ideal in 'Time'), personal judgment, experience or opinion have no place in scientific journals, while in general magazines they abound.
- scientific knowledge comprises generalisations which are valid and reproducible under well defined circumstances by observers in any country, social or ethnic group. General magazines often discuss issues of special interest to a particular social group or to a specific country (e.g. to beauty and poise in 'Time').
- there is little discussion of ethical questions in science journals (what ought to be done). ST generates new ethical problems, it can assess risks associated with new discoveries, but ethical questions cannot be resolved by the application of scientific method alone. In general magazines ethical, business and political questions and opinions are common (see 'Time' covering political figure of Stalin)
- the language used in scientific journals is very specialised and knowledge is efficiently presented (e.g. by using special symbols and equations) as the cover page of "Journal of Physical Chemistry" reveals. This language is often incomprehensible without professional training. This fact may lead to misunderstanding by non-scientists. For example, the 'Theory of Everything' would represent the theory which unites description of four fundamental physical forces and not a theory that would explain everything we would like to know. In general magazines, on the other hand the reported information is easy to follow (cover of 'Time' shows Anita Colby who was well known from advertising, especially cigarettes).
- methods used for gathering information in Science are restricted – only data obtained and analysed by scientific method are reported in scientific journals. Information gathered must satisfy

criteria for quantification, use of appropriate instrumentation, reproducibility, careful sample selection. In the field of general knowledge, e.g. journalism, this is not the case. For example, many surveys which are published in newspapers do not comply with such criteria. The surveys do not provide details of the principles which were applied when, for example, samples or people were selected for a survey.

- scientific publications often carry multiple authors, unlike novels and newspaper articles. Scientific articles also carry the date of receipt of the manuscript in order to establish scientists' priority claims; this does not apply to articles in general magazines.
- the knowledge described in scientific publishing aims to be the most accurate and reliable knowledge about Nature available at the time of publication. General magazines and media often report speculations and opinions. In order to ensure accuracy and depth of insight, scientists have developed, through experience, the 'scientific method' which stipulates how scientific knowledge should be acquired. Scientific method is a set of procedures, norms and processes which ensures (as much as possible) that errors are avoided. This reliability is achieved through specific methods used by individual scientists and through the scrutiny of their results by the scientific community (see Chapter 2). General magazines on the other hand often report rumours or unconfirmed information subject to minimal editorial scrutiny and are governed by market demands. They provide information which readers wish to read about irrespective of its accuracy or information which they find spectacular or intriguing. For example, the notorious J. Stalin was "Man of the Year" in 1940 issue of Time magazine (see the cover which follows)!

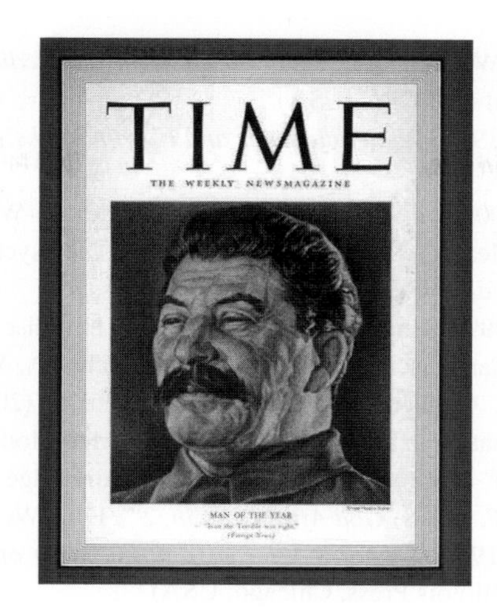

Bibliography for Chapter 1:

1. a) Mackay, A.L. (1991) *A Dictionary of Scientific Quotations*, (IOP Publishing, UK).
   b) Gaither, C.C., Cavazos-Gaither, A.E. (2000) *Scientifically Speaking*, (IOP Publishing, UK).
2. a) Kingsnorth, P. (2009) A Windfarm is not the answer, *Guardian Newspaper*, (July 31, UK).
   b) Monbiot, G. (2009) This is bigger than climate change. It is a battle to redefine humanity, *Guardian Newspaper*, (Dec 14, UK).
   c) Bunting, M. (2010) To tackle the last decades' myths, we must dust off the big moral questions, *Guardian Newspaper*, (Feb 21, UK).
3. http://www.nsf.gov/news/speeches/colwell
4. a) Cassidy, R. I. (2002). Do scientists understand science?, Can.Chem.News, Sept issue, pp. 37-38.
   b) Crease, R.P. (2000). A top ten for science and society, Phys.World, Dec issue, pp.17-18.
5. a) Lall, V. (2007). Next steps for physics graduates, Phys.World, Oct issue, pp.48-49
   b) O'Byrne, J., Mendez, A., http://science.uniserve.edu.au/pubs/procs/2008/199.pdf.
6. Devlin, K. (1998). Rather Than Scientific Literacy, Colleges Should Teach Scientific Awareness, Am.J.Phys., 66, pp.559-560.
7. Ball, P. (2002). Beyond words: science and visual theatre, Interdiscip.Sci.Rev. 27, pp.169-172.
8. Frayn, M. (2000) *Copenhagen*, (Anchor Books, USA).

9. Weisstein, E.W., (2003) *CRC Concise Encyclopaedia of Mathematics*, 2nd ed., (Chapman and Hall/CRC, USA).
10. Derry, G.N., (1999) *What Science is and How it Works*, (Princeton University Press, USA).
11. Lee, J.A., (2000) *The Scientific Endeavour*, (Addison-Wesley Longman, USA).
12. Vohs, K.D., Mead, N.L., Goode, M.R. (2006). The psychological consequences of money, Science, 314, pp.1154-1156. (17 Nov issue)
13. Ingold, T., (1999). Human nature and science, Interdiscip.Sci.Rev. 24, pp.250-254.
14. Strippoli, P., Canaider, S., Noferini, F., D'Addabbo, P., Vitale, L., Facchin, F., Lenzi, L., Casadei, R., Carinci, P., Zannotti, M., Frabetti, F. (2005). Uncertainty principle of Genetic information in a living cell, Theor.Biol.Med.Model. 2, pp.40-45.
15. a) Haught, J.F. (2006) *Is Nature Enough?*, (Cambridge University Press, UK).
    b) Haught, J.F. (2008) *God After Darwin*, 2nd. Ed., (Westview Press, USA)
16. Bauer, H.H. (1992) *Scientific Literacy and the Myth of the Scientific Method*, (The University of Illinois Press, Chicago, USA)

Chapter 2

Philosophical & Methodological Aspects of Science

<u>Some fundamental questions considered in this chapter:</u>

How does Science operate and why it operates in a particular way?

How can we use and understand scientific evidence?

How do we acquire knowledge and what is the meaning of "understanding" and "knowing"?

How does one think in Science?

What confidence can we have in scientific methods, results and their interpretation?

Why should Nature be amenable to human understanding at all?

All citations were taken from ref [1]. Further discussion of topics in this Chapter can be found in refs. [2-7].

Science is a human activity. Humans are social beings, and so Science can also be expected to be a social, communal activity. Science is performed in the community of scientists. Let us discuss this community first.

Scientific community (SC)

Science is performed within the scientific community (SC). The SC ensures progress, self-correction and free exchange of ideas/data. The SC is small it comprises less than 1% of the World population. SC is a group of people who share common views about the nature of the Universe, the use of scientific method, and the aims and objectives of Science.

SC members have different roles: some are creative, original and independent; they break new grounds. Others are systematic, predictable and persistent; they gather essential information on which new ideas/insights can be based. Both types of scientists are necessary for scientific progress, but as in most communities, friction between them may also arise from time to time.

Example of the Scientific Community's role

The important role of the scientific community in the discovery of DNA

Watson and Crick used crucial data obtained by other scientists (such as Franklin and Chargaff, both pictured below) in an imaginative, original way to come up with the double helix model which explains DNA function.

Erwin Chargaff (1929-1992) showed that in DNA the number of guanine bases equals the number of cytosine bases and that the number of adenine units equals the number of thymine units. This strongly hinted at the base pair makeup of the DNA molecule.

source: http://qspace.library.queensu.ca/html/1974/136/bioinfor.htm

Rosalind Franklin (1920-1958) studied the structure of DNA by X-ray crystallography and this guided Watson and Crick in developing models consistent with the X-ray data.

source: http://www.silobreaker.com/today-in-history-rosalind-franklin-and-the-discovery-of-dna-the-primate-diaries-5_2262482090819846144

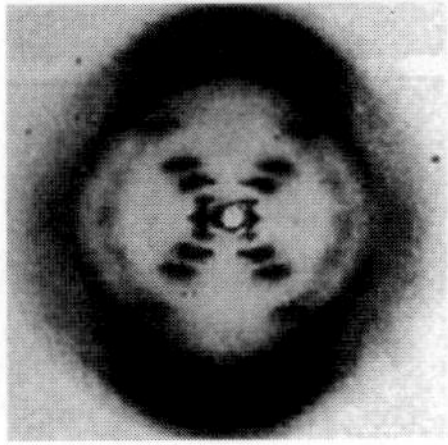

Rosalind Franklin's X-ray diffraction photograph of DNA (1953).

source: http://genome.jgi-psf.org/Chr19ncbi34/Chr19ncbi34.home.html

Watson and Crick with their mechanical model of the DNA molecule.

source: http://www.odec.ca/projects/2007/knig7d2/Scientists.html

Without these various scientists exchanging ideas within the scientific community, DNA discoveries would not have been possible. This example and the citation below show that Science is a cumulative activity in which scientists build on the results obtained by their predecessors.

> Dwarfs on the shoulders of giants see further than the giants themselves.
>
> **Stella Disacus**

Members of the scientific community are at various times engaged in various types of scientific activities:

- gathering data;
- consolidation (application of law/theory in established areas);
- extension (application of law/theory to new areas);
- reformulation (clarification, simplification, improvement of theories);
- theory construction (creation of new laws/theories).

Two important mechanisms – peer review (PR) and the reward system (RS) operate within the scientific community to guide and stimulate research activities.

Peer review and reward systems

Science aims to provide the most accurate possible description of aspects of Reality. The PR system is therefore methodologically crucial for Science, because it greatly helps in (although it cannot guarantee) the elimination of erroneous results/conclusions. PR is also applied to research grant applications in order to minimize waste of research funding.

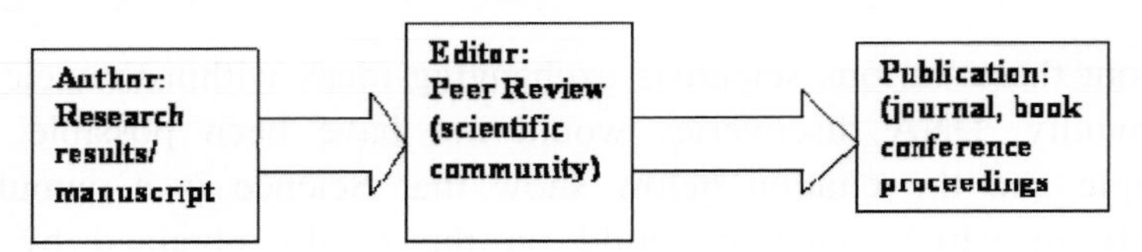

The peer review system

PR means that scientists working in a specific area are asked to review the work of their fellow scientists. This happens when e.g. a scientist submits manuscript to science journal or when he/she applies for research funding. For example, PR may be expressed through the following questions put to the scientific reviewer by the journal editor regarding a fellow scientist's work. (The work may comprise manuscript submitted for publication to a journal or a research grant proposal submitted to a funding agency):

- Was the research work performed by appropriate methods?
- Are the conclusions arrived at justified by the data obtained?
- Is the contribution worthy of publishing (is it novel, interesting for readers)?

Peer review is anonymous. It is not foolproof, but is the best current mechanism for ensuring 'quality control' in Science. Some common problems with the PR system include:

- reviewers may be competitors or collaborators and hence too harsh or too lenient in their assessments of the submitted work
- PR may reject ideas which contradict prevailing theories

- PR may be lax due to the large number of scientific contributions that reviewers have to assess.
- PR system has another potential flaw. Scientist may review someone else's work favourably in order to receive a favourable review of his own work in return. The anonymity of PR system tries to prevent that. However, the fragmentation of Science into narrow, specialized fields of research may sometimes allow one to make an informed guess about the identity of the anonymous referee.

The current practice which requires the authors of the submitted manuscript to nominate potential referees may also weaken the PR system. A better practice to implement would be to remove authors' names and affiliations from the submitted manuscript and thus preclude reviewers from identifying the authors. The principal cause of problems for PR system is the significant increase in the number of manuscripts and research projects which are submitted to journals and funding agencies. The increase is due to social and institutional pressures on scientists to continuously produce more and better scientific results. This tends to overload the PR system and allows some unsatisfactory research to slip through the PR system. The present comments also highlight the fact that external (socioeconomic) factors have some influence on the scientific practice.

The rewards for scientist's work may come in many forms, but for most scientists, peer recognition is the most important one. Peer recognition may be expressed as:

- having a discovery named after the scientist (*eponymy*) e.g. Boyle law;
- award of prizes (e.g. Nobel Prize) and election to memberships of prestigious scientific societies (e.g. Fellow of the Royal Society in UK; which entitles the member to add FRS title to his name);
- opportunity to work at prominent institutions/universities and promotion within the University hierarchy (assistant professor→ associate professor→ full professor). The promotion is often

based on the number of publications and the amount of research funding obtained by the individual scientist;

- large number of citations of one's published work by fellow scientists.

The training of scientists is lengthy and specialized. Young scientists learn not only about how to select and study research problems, but also about how Science works, how to interact with other scientists, how to write articles, reports, manuscripts, grant proposals and so on.

Typical career stages in becoming a scientist comprise:

- college/university undergraduate (engaged in 'textbook science', learns experimental and communication skills)
- PhD student/postgraduate student (research in 'frontier science', specialisation, performs original research, supervised)
- postdoctoral work (performs semi-independent research, unsupervised)
- academic staff (performs independent research leading to tenure, promotion)
- contract researcher (company consultant, industry researcher).

Routes to scientific discovery and selection of research problem [2,8]

There are no 'prescribed' rules for making discovery and achieving scientific success. This is what makes Science a rewarding career, difficult, but exciting! However, there are established principles for evaluating ideas, measurements, observations, models. Some possible routes to discovery are listed here and discussed in more detail in ref.[2], pp.11-51.

A. Serendipity and methodical follow up: For example, the well known Fleming's serendipitous discovery of the species of penicillin fungi with antimicrobial properties. This route **requires** systematic follow up of the unexpected result or idea. This is a risky decision to make because the supposed lead may prove to be unfruitful after all. However, without such risk there is no discovery.

B. Imaginative thinking – 'Eureka' intuition: For example, Kekule's postulated structure of benzene molecule which ran contrary to the ideas on chemical bonding during his time.

C. Theoretical (mathematical) modeling: For example, the discovery of the band structure of solids shows how theoretical model based on fundamental principles leads to predictions of previously unknown properties.

D. Exploration and observation: For example, Humboldt's discovery of biogeography.

E. Logical analysis (deductive and inductive): For example, Jenner's discovery of smallpox vaccine which combined observation, experiment and deductive reasoning.

F. Pattern recognition: For example, Mendeleev's discovery of Periodic Table. See Case Study 2 in Chapter 4 for details.

G. Use of new instrumentation: For example, the discovery of quasars and microorganisms. Neither of these discoveries could have been made without new equipment which became available shortly before these discoveries made.

H. Study of discrepancies/anomalies: For example, the discovery of rare gases by Ramsey who observed only a small discrepancy in measured vs. calculated air density. He decided that discrepancy was not due to experimental measurement error, but had a deeper significance.

Reflection:

1. Select two discoveries from the above list and describe which of the components (A-H) were involved in them.

2. Find origin of the word 'eureka' and explain how this word relates to the discovery of benzene structure.

Tips: see ref. [2], pp.11-25 for discoveries and English Dictionary for 'eureka'

The discovery in Science is not purely accidental, the scientist needs to be prepared to spot and identify an interesting problem and possible solution as the quotations below indicate.

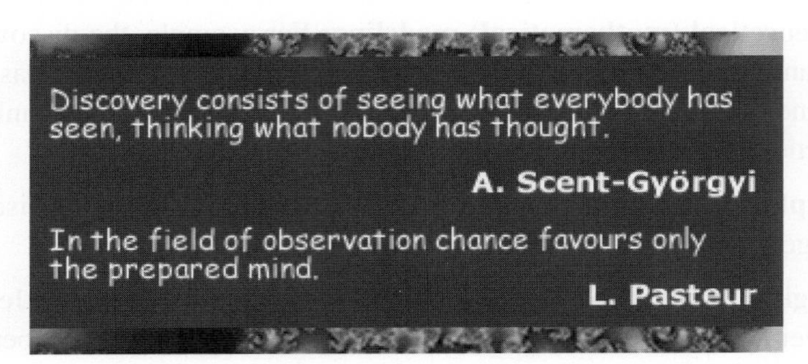

Discovery consists of seeing what everybody has seen, thinking what nobody has thought.
A. Scent-Györgyi
In the field of observation chance favours only the prepared mind.
L. Pasteur

The selection of a research problem by the individual scientist is governed by several factors and this selection influences the final outcome of the research project:

- personal interest;
- technical expertise and material resources available (equipment);
- the availability of funding for the project.

General characteristics of Science

We list some general characteristics of Science (S) which are useful when we wish to assess whether certain activity should be considered scientific or not.

- S is a form of knowledge and a type of activity;
- S is guided by the laws of Nature;
- S is explanatory with reference to the laws of Nature;
- scientific results are verifiable/falsifiable by using empirical procedures;

- S is open-ended, i.e. its conclusions are tentative (this is shown by changes in theories/ methods through time).
- S is performed within a scientific community which is global in its profile
- S is based on critical thinking and uses well-defined types of logical reasoning;
- S assumes that there is an order in the World which is intelligible to humans

> This coherence of valuation throughout the whole range of science underlies the unity of science. It means that any statement recognized as valid in one part of science can, in general, be considered as underwritten by all scientists. It also results in a general homogeneity of and a mutual respect between all kinds of scientists, by virtue of which science forms an organic unity.
>
> **M. Polanyi**

There is no single, general scientific method which, when applied to data and observations, will produce Science as discussed by Bauer [8]. However, this does not imply that any form of activity can produce scientific results. The two important principles of scientific method are:

- Scientific method is used to study Nature and may not be appropriate for studying other problems and questions e.g. related to ethics, beauty, emotions etc.
- The results obtained by scientific method are based on the combination of empirical evidence and logical reasoning and are valid only as long as they are consistent with general principles of the natural world. Conclusions and reasoning which contradict empirical evidence may sometimes be expressed through our emotions and embedded in artistic and religious creativity. This

comment is not meant to denigrate other types of creativity which after all greatly enrich our human character, but only to point out that such creativities do not and cannot use scientific method. Science serves as a powerful social correcting influence by making us aware of the fact that we are part of Nature and subject to its laws and constraints. Science is useful in dispelling myths, fantasies, wishful thinking and irrational behaviour. Acting as a check on our instinctive desires does not make Science popular, but such 'reality therapy' is surely useful.

Reflection:

Why does Science operate in a different way from Art? (Refer to Chapter 1)

What is the public misconception of the role of 'genius' in Science?

Tips: consider the relationships between Art and Nature and Science and Nature. Consider whether 'genius' as a person in Science is unique.

The cycle of scientific activities

Science is not performed as a sequence of preordained steps nor as a single step, but as a cycle of activities. Different sciences put different emphasis on different elements of the cycle, and can be classified as:

- **observational** (astronomy, geology, biology- use observation because in these fields experiments are difficult to perform)
- **experimental** (physics, chemistry, biology-predominantly use experiment)
- **directional** (biology, geology, astronomy-they study large, slowly and irreversibly evolving systems and processes)
- **elemental** (chemistry, physics-investigate the most fundamental processes)

- **data-rich** (chemistry, biology-the amount of empirical data collected is greater than the extant theories can encompass and explain)
- **data-poor** (astronomy – relies on observation. The amount of data acquired is small compared to predictions deduced from cosmological theories. Such theories are often speculative and their predictions reach beyond our current experimental capabilities e.g. Black Holes, Big Bang etc).

The concept map which follows shows the elements (activities) in the cycle.

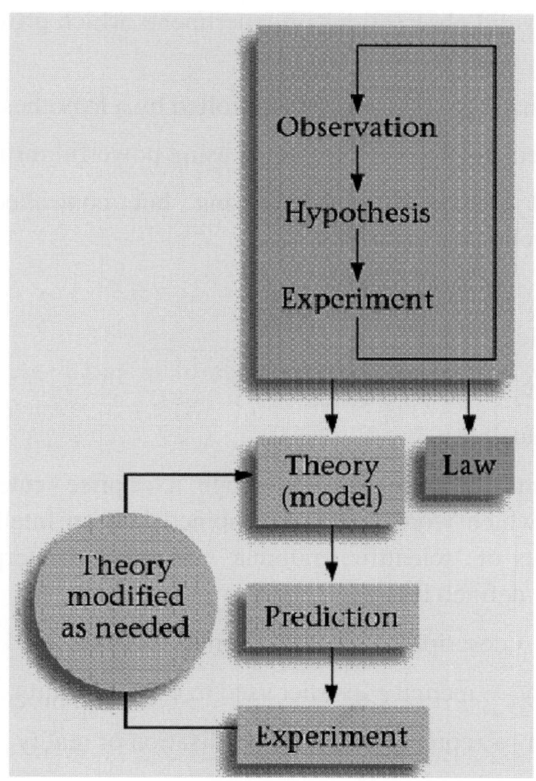

Concept map of the cycle of scientific activities

Scientific language

The cycle of scientific activities, comprises several entities and processes which are defined below. These entities and processes can be considered elements of 'scientific language'. Observations and experiments provide facts and data. Facts and data have the role of words in the 'scientific language'.

- Facts → events, states
- Data → symbolic representations of facts (e.g. numerical records, graphs etc.)

Scientifically valid observations or experiments which provide data must be:

- systematic (organized and controlled by a hypothesis or theory)
- detailed and accurate (obtained using powerful instrumentation)
- varied (obtained under varying but controlled conditions; 'controlled experiment')

Scientifically valid data must be:

- reliable,
- precise
- reproducible

Other elements of scientific language comprise rules (grammar) according to which words can be combined into an intelligible whole. These elements of 'scientific grammar' are used to interpret facts and data. They are defined below:

- Law → descriptive generalisation of reality
- Theory → mental construct used to explain reality
- Model → representation or visualisation of reality

- Hypothesis → working conjecture about reality (hypothesis may become a theory after undergoing appropriate scientific testing via experiment or observation)

Metaphors

Another element of scientific language is a metaphor. This element is widely used in other fields of human activity besides Science e.g. in literature. A metaphor is a representation or visualization of reality. In other words, it is a statement that two things are the same when in fact they are only similar in some aspects. Metaphor helps to implant knowledge about something abstract/unfamiliar by utilizing the knowledge/experience of something concrete/familiar. Metaphors are necessary when describing complex events or objects which are beyond direct human sensory experience.

Metaphors communicate only **a part** of the complex reality we are trying to describe, so sometimes a string of metaphors may be necessary. For this reason, scientific metaphors **must not** be taken literally. For example, a 'black hole' isn't like a hole found on road surfaces! It's a metaphor.

Examples of scientific statements (sentences) with scientific metaphors (underlined):

Statement: Metals have structures of close-packed spheres.

Metaphor implies: Metal atoms are hard spheres.

Statement: A radioactive nucleus decays to the nucleus of another atom.

Metaphor implies: The nucleus is a dead plant or animal.

Atoms are not hard spheres nor are nuclei dead organisms. As noted, metaphors should not be taken too literally.

Laws of Nature

The origin and practice of Science is based on the assumption that there is a hidden order in Nature, which can be uncovered by empirical investigations. This hidden order could be expressed succinctly in the form of fundamental and general principles, or laws of Nature. The laws can be coded in mathematical form for easy manipulation and application. Consider the example of the boiling of water (diagram follows) in which the apparently chaotic motion of individual molecules is described via mechanical principles which include velocity and momentum.

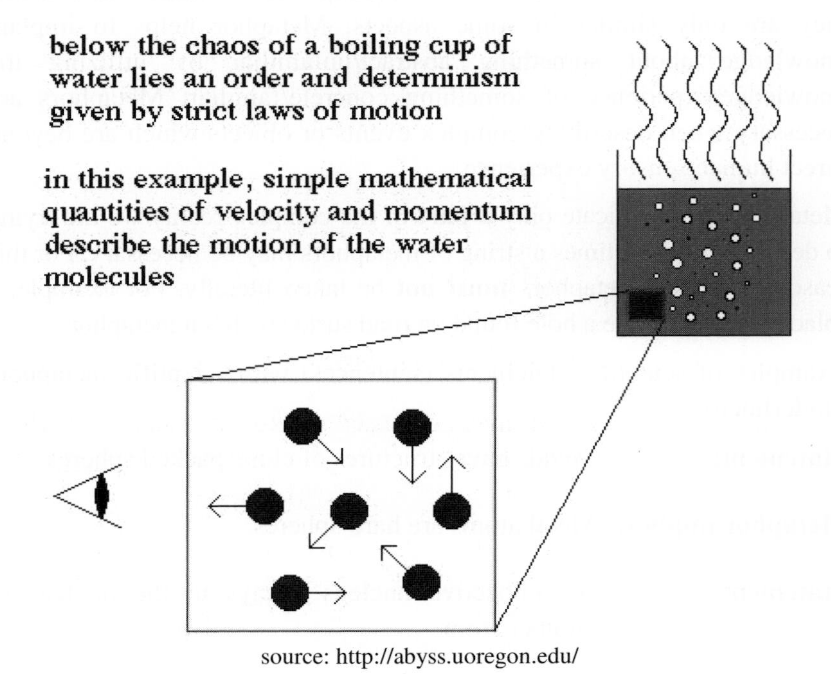

below the chaos of a boiling cup of water lies an order and determinism given by strict laws of motion
in this example, simple mathematical quantities of velocity and momentum describe the motion of the water molecules

source: http://abyss.uoregon.edu/

In Science, the processes of examination, observation or experimentation are used to reveal meaningful connections between diverse events or entities. Direct connections are usually apparent to unaided senses; however, underlying causes associated with the laws of Nature are not readily apparent. The observations of phenomena themselves may not be readily intelligible or understandable. We often use abstract theoretical frameworks as basis for explanations and to serve as guides for directing and organizing experimental work. These theoretical frameworks are called scientific theories; they integrate and organize data into an intelligible whole.

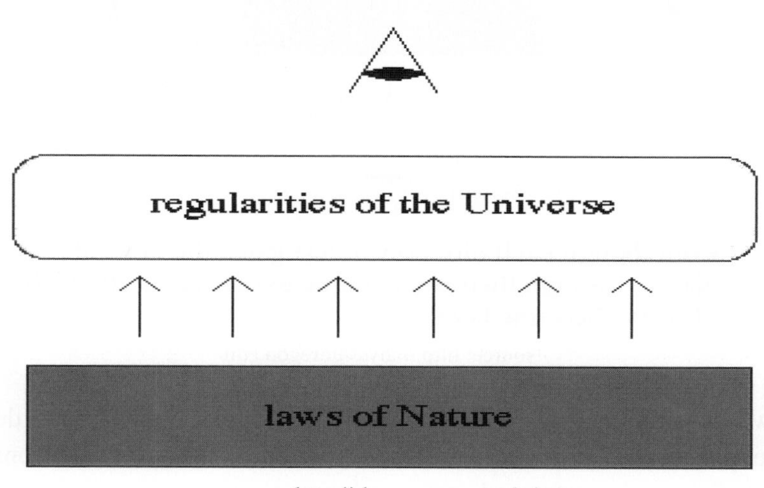

source: http://abyss.uoregon.edu/

Laws of Nature are human inventions used to express regularities which we observe in the Universe. While laws are human inventions, the regularities have independent existence in the physical world.

The laws of Nature are human attempts to capture regularities and patterns of the World in a systematic and concise way. The existence of regularities is supported by observations. We do not impose regularities on Nature, but only describe them in the guise of natural laws. While the forms of the laws are human inventions, laws reflect, albeit imperfectly, real events found in Nature. It is the invariance and general validity of laws of Nature which underpins the meaningfulness of scientific activities and assures their success.

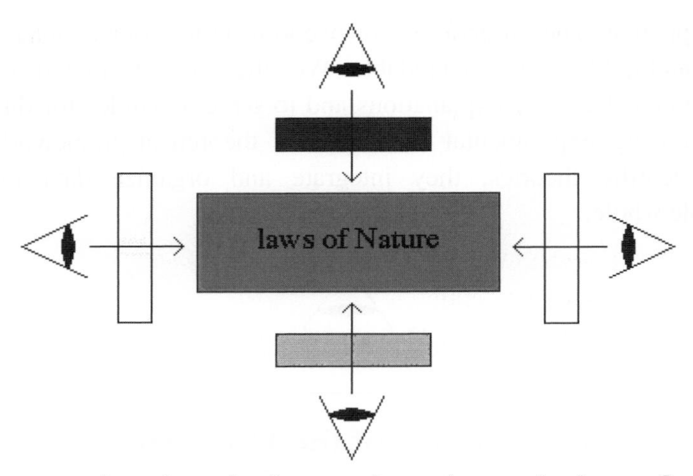

even though each observer investigates the laws of Nature through their own unique experiments, they all observe the same laws

source: http://abyss.uoregon.edu/

New laws often lead to new discoveries. Laws are independent of design of experiments, of observers (scientists) or the cultural environment where Science is performed.

Fundamental laws of Nature describe deep connections between different physical processes and physical entities (objects). When a new law is developed, it is tested in different contexts, which often leads to the discovery of new, unexpected phenomena. The predictive power of scientific law demonstrates that we are dealing with real regularities in Nature and not imposing our own perceptions onto natural processes.

The laws of Nature have independent existence outside physical set up of an experiment. Success of the scientific method rests on the reproducibility of empirical results. An experiment is repeated and the same laws of Nature apply, but the initial conditions of the experiment may be varied. There is a clear functional separation between laws and initial conditions. Laws are invariant, initial conditions are not. For example, the initial positions from which my car starts moving can vary widely, but the laws of motion to which my car is subject during the motion do not!

If Shakespeare, Beethoven, or van Gogh had not been born it is unlikely that anyone else would have ever created the works of art identical to theirs. But is this true for scientists? Would someone else have discovered classical laws of motion if there had been no Newton? Most probably someone else would, because the regularities in Nature exist independently of the ability of individuals to perceive them. Works of Art on the other hand bear indelible stamp of their creators (artists). This highlights the fact that Science is a collective and cumulative activity while Art is an individual, non-cumulative activity.

From the point of view of cumulative property, Science progresses Art does not. Relativity theory is a progress vs. Newton laws of motion. Modern Art is not a 'progress' vs. Renaissance Art, the two artistic traditions are just different.

The solution of a scientific problem must satisfy exacting criteria and demands summarized below:

- Events are studied by observations and experiments. They must be combined into an organised pattern.
- Scientific laws:
    - describe classes of events, not single events;
    - they show functional relationship between two or more events;
    - they are supported by large amount of supporting experimental evidence (with little disconfirming evidence/exceptions); and
    - laws do not explain why certain things happen, e.g. Newton laws do not explain what gravitation is, only how it works!

Models and theories

Science uses models and theories to describe the physical world. Data are interpreted and organized through theories. Scientific theories can be considered similar to models of the real world (or parts of it) and the vocabulary of Science often manipulates models rather than real objects. Sometimes when the term `discover' is used in a scientific model or theory it refers to establishment of a mathematical relationship that has been revealed. A true discovery would refer to direct observation of the phenomenon in Nature which is not easy for many objects or phenomena (no one has yet directly observed a 'black hole').

Theories relate and explain laws in terms of the fundamental framework and concepts which were not used in the derivation of laws.

Reflection:

Why we cannot use the same elements to build the law and theory?

Tip: consider if this would lead to a logical fallacy; also see ref. [9], p.140

Models are simplified descriptions of reality. They are used, because real objects may be too large (e.g. planets), too far away (e.g. galaxies), too complicated or ethically unsuitable for experimental study (e.g. use of humans for studying effects of toxins). Some models are accurate, the others less so. The accuracy of the model depends on the extent of our understanding of the phenomenon being studied.

Models can be of hardware (e.g. DNA model used by Watson and Crick) or of software type (mathematical algorithms) and can be used to make predictions about events. The crucial thing when using a model is to understand how the model differs from reality i.e. what approximations and assumptions were made in constructing the model. This will determine how much we can rely on the results derived from the model. For example, animals are used to model biochemistry of human bodies in medical research, but we need to assess the transferability of the results from animal studies to humans. Animals and humans have similarities, but also differences in biochemical processes in their bodies.

There are always several possible models for the same reality; **models are not unique**! This is because every model is a representation and a subset of reality, not reality itself! We obtain a model (a subset) by discarding one or more elements of reality (the parent set). Which elements do we discard? It depends on the problem we are studying and on our judgment as to which elements are essential and which are not. The Venn diagram below illustrates this point.

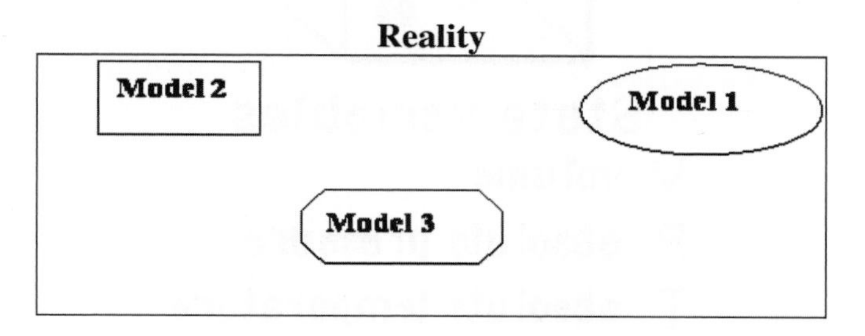

The diagram above shows how several models (various smaller shapes) can be formed from the same reality (large rectangle)

The non-uniqueness of models is often the source of frustration for students in scientific disciplines, because students have the misconception that scientific knowledge of a specific phenomenon must be complete and clear cut. General public often shares the same misconception and tends to interpret the profusion of models in Science as a sign of ignorance or weakness.

Let us look next at an example of how law, model and theory are connected.

An example of a scientific theory:

Kinetic theory of gases

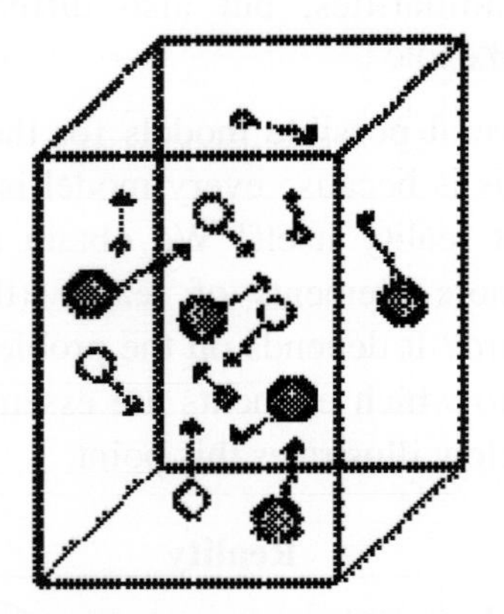

State variables

V volume

P absolute pressure

T absolute temperature

Many observations were made about how the pressure of gas (p) changes with volume (V). They are facts. The facts can be summarized into a law (Boyle law):

'the smaller the volume of the gas the higher its pressure'.

However, this law does not explain why this is so.

Kinetic theory of gases uses the model of atoms and molecules in the gas as small hard balls which move and collide ceaselessly according to Newton's laws (note that kinetic theory does not use p, V in its explanation). Kinetic theory explains Boyle law (pV=k) as follows. The pressure arises from collisions between particles and the wall of the container; the smaller the volume the more often particles would collide with the wall and thus we would register higher pressure. Using kinetic theory of gases in its mathematical form (i.e. as a mathematical model of gas) we can derive the ideal gas law: **pV = k** (pressure is inversely proportional to volume at constant temperature). The law derived from kinetic theory is (and must be) consistent with the experimentally derived law ('the smaller the volume the higher the pressure').

$$\mathbf{pV = k}$$

(i.e. pressure is inversely proportional to volume at constant temperature).

The law, pV = k, allows us to make predictions about states of gas at any pressure and volume. The predictions can be tested and this law can thus be experimentally verified.

The relationship between Nature (reality), models, theories and experiment is shown on the diagram below.

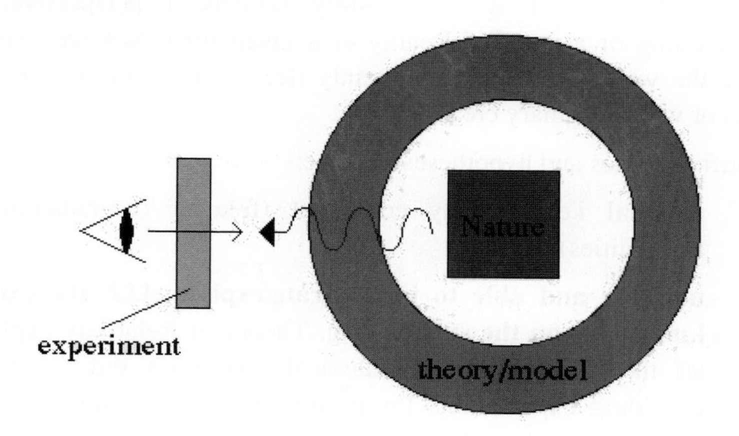

source: http://abyss.uoregon.edu/

The relationship between a theory or a model and the real system presents an important problem. For example, how do we know when a model is merely a computational device and when it does actually describe reality?

Scientific theories are descriptions of Reality; they do not constitute that Reality. As long as a theory adheres to evidence which is a reliable guide, there is confidence that theory is a faithful representation of

Reality. However, some advanced theories in modern physics break this boundary (for example, they use the notion of 'virtual particles' in quantum physics). 'Virtual particles' are never directly observed, but are used as a convenient, simple way of describing inter-particle interactions.

Models, theories and methods that are broad and encompass significant portions of many scientific fields are called **paradigms**.

Reductionism is one of the founding paradigms of Science, but it does not provide a complete description of Nature. However, 300 years of scientific progress which relied on reductionism proves that reductionism is a paradigm well suited to study of Nature. A particular paradigm is neither right nor wrong, but merely reflects a perspective, our understanding of aspects of Reality at a given time. Science may not deliver the whole truth, but it certainly deals with truth and not with dogma or with imaginary creations.

Scientific theories and hypotheses must be:

- **rational** i.e. logically consistent (free of contradictions or ambiguities);
- **unifying and able to incorporate/explain *ALL* the existing knowledge on the subject** (e.g. Theory of Relativity explained all the processes which classical mechanics did as well as providing explanations for phenomena at very high velocities which classical mechanics could not explain or predict)
- **extensible** i.e. theories must not explain only what is known, but also predict what is unknown e.g. the Periodic Table devised by Mendeleev predicted the properties of unknown chemical elements before they were discovered. Theories should preferably have wide scope and explanatory power. Good theory explains a large body of facts; a theory should extend beyond particular laws or sub-theories that it was created to explain;

- **testable/verifiable by experiment/observation** (predictions made by the theory should agree with the results of experiment /observation);
- **relevant** (i.e. clearly related to the physical World and not to imaginary objects);
- **simple** (theory should incorporate as few assumptions as possible and bring order into the phenomena that would be considered isolated without it; this property of simplicity is related to 'inference to the best explanation' and Ockham's razor principle).

The second, third and fourth requirements (unifying, extensible and testable) are sometimes not satisfied by theories in social sciences which prompted an opinion that social sciences are not 'true sciences'.

Distinguishing between Theory and Truth

Can scientific theory ever become Truth? No! Let's look at the fundamental explanation of why this is so. Mathematician *Gödel* has proved a theorem ('incompleteness theorem') which deals with mathematical axiomatic systems and defines ultimate limits of logic. The theorem states that in any axiomatic system there exist propositions (theorems, ideas, statements, conclusions) which are undecideable i.e. whose truth or falsity cannot be proved or disproved from within that same system. In other words, for every formal (logical) system there are limits on the amount of 'truth' that we can squeeze out of it. There are statements deduced from within the system which cannot be proven true or false without enlarging the system itself.

Gödel theorem is fundamental to mathematical logic and caused quite a stir when it first appeared. As one can see from the Figure 6, Gödel's personal life was also extraordinary. He was one of most important mathematical thinkers of the 20th century. The photo below shows him in the company of another scientist who has profoundly influenced modern science, Einstein.

Kurt Gödel

- Born 1906 in Brno (now Czech Republic, then Austria-Hungary)
- 1931: publishes *Über formal unentscheidbare Sätze der Principia Mathematica und verwandter Systeme* (On *Formally Undecidable Propositions of Principia Mathematica and Related Systems*)

- 1939: flees Vienna
- Institute for Advanced Study, Princeton
- Died in 1978 – convinced everything was poisoned and refused to eat

source: http://en.wikipedia.org/wiki/Kurt_G%C3%B6del

Fig. 6.

Reflection:

Epimenides (the Cretan) stated: 'All Cretans are liars'. Is this statement true, false or undecideable? Propose the modification of the statement which will avoid the paradox!

Tips: Firstly, consider the claim to be true

(Was Epimenides lying then?).

Secondly, consider the claim to be false (Was Epimenides then telling the truth?)

To remove the paradox, try modifying the phrase "All Cretans….."

Possible methodological errors in Science

Science is a human activity and humans are prone to making errors. Science is therefore not immune from errors. The following list contains some errors and pitfalls which pertain to scientific practice. Knowing what they are is useful so that we can avoid them.

Ptolemaic method → uses a theory with too many adjustable parameters. The theory which contains a multitude of adjustable parameters can be tailored to any desired outcome. Such theory is obviously useless.

Platonism → the claim that hypothesis must be true because it is convincing, simple and elegant. Simplicity, elegance and persuading power are no substitute for Truth. Hypothesis needs to be consistent with other, existing knowledge. Simplicity, elegance and persuading power are subjective criteria used in arts, marketing or politics, but not in Science! However, simplicity may be a useful criterion as we shall see later (Ockham's razor principle).

Experimenter's regress → circular reasoning fallacy

- I want to detect a hypothetical quantity A
- I build a detector D
- Is D satisfactory?
- I must check and see if D can detect A

I cannot test D if I do not know (independently) whether A exists. A quality test of the experimental setup must be made whose validity is independent of the experimental setup being tested.

Reflection:

Select the best answer giving the proper order of steps in the scientific process is:

A. predictions experiment observation hypothesis

B. experiment observation predictions hypothesis

C. hypothesis observation experiment predictions

D. observation hypothesis predictions experiment

E. None of the above

Tip: try D as an answer and compare it with the cycle of scientific activities

Thought processes and logical reasoning in Science

General or universal statements

Scientific knowledge invariably involves general statements such as:

- metals expand when heated,
- acid turns litmus paper red.

These statements refer to all events of a particular kind, that is:

- to all instances of metals being heated, and
- to all instances of litmus paper turning red in acid.

But what about the reverse type of statement? What about individual observation statements that constitute facts? Such statements are called singular statements.

Singular statements

Singular statements are specific claims made about a state of affairs at a particular time, i.e.

- the length of this copper bar increased when heated in the fire yesterday
- when this rod of iron was cooled this morning it got shorter.

If we have a large number of such observation statements, how can we combine them into general/universal scientific laws? How can we go from many **singular** statements to a single **universal** statement? We shall discuss how to do that next as part of discussion on scientific arguments.

Types of scientific arguments

We briefly describe main types of scientific arguments and modes of reasoning [9].

1. Retroduction: anomaly → hypothesis

Example: The orbit of Mars was not consistent with circular motion, hence Kepler hypothesized that planetary orbits are elliptical.

2. Deduction: general principles → prediction

Example: Starting from the notion that there is no absolute reference frame in the Universe and that physical laws should be invariant with respect to different frames of motion, Einstein deduced the Theory of Relativity.

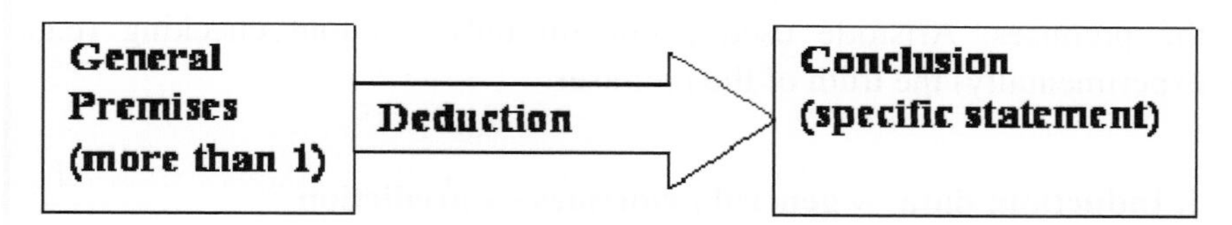

Let us look next at some examples of deductive reasoning.

(1) All books on Science are boring	(True)
(2) This is a book on Science	(True)
(3) This book is boring	(True)

Thus, deduction is truth-preserving. If premises (1) and (2) are true **and** if the logical argument is valid then conclusion (3) must necessarily be true. However, if the logical argument is invalid or if premises are untrue, the conclusion may or may not be true. In Science we are only interested in conclusions which **must** be true hence the deductive logical reasoning is important. In the next example the logical argument is invalid (*many*, not *all* books are postulated)

(1) Many books on Science are boring	(True)
(2) This is a book on Science	(True)
(3) This book is boring	(False)

It is also possible for the logical argument to be valid, but if one of the premises is false the conclusion may also be false.

(1) All dogs have 5 legs	(False)
(2) Bobby is my dog	(True)
(3) Bobby has 5 legs	(False)

Deduction cannot establish **new truths** i.e. it cannot tell if premises are true or not. Deduction only reveals the information already contained in the premises. Aristotle used deduction only, without checking (e.g. experimentally) the truth of the premises.

3. Induction: data → general principles → prediction

F. Bacon in the 16th century proposed the use of 'inverted' logical method: induction. In induction method the flow of reasoning is as follows:

- some statements → all statements
- singular statements → universal statements

For example, studying the ratios of volumes in which **many** simple gases combine and react, Gay-Lussac postulated that **all** chemical compounds are formed by combinations of elements in integer ratios e.g. 2 mass units of hydrogen will combine with 16 mass units of oxygen to give 18 mass units of water.

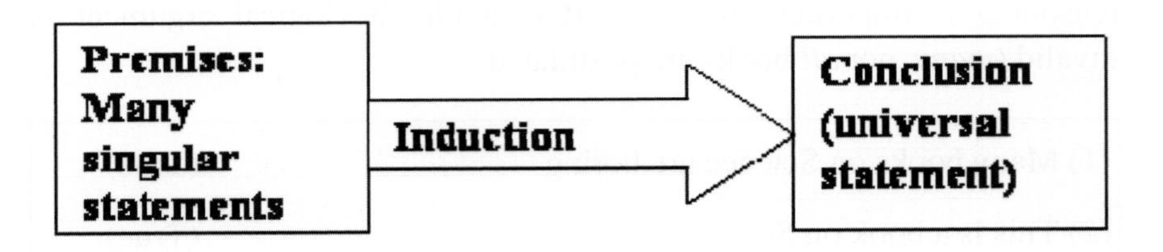

Another example of induction is:

(1) Metal *a* expands when heated on occasion t1
(2) Metal *b* expands when heated on occasion t2
(3) Metal *c* expands when heated on occasion t3

(n) Metal *x* expands when heated on occasion tn

Therefore we conclude that all metals expand when heated.

The problem with induction (**Hume problem**) is that if some (or even many) statements about events of a particular kind are true there is no necessity that all such statements be true. The more events we have observed, the more probably correct the inductive conclusion becomes, but it still does not guarantee that it is always true.

Inductive inference presupposes the 'uniformity of nature' i.e. that the objects we have not examined will be similar and behave similarly to those objects of the same kind which we have already examined. Inductive conclusion assumes that if statements were true before, they shall always be true. Yet consider Bertrand Russell's 'inductionist' turkey as an example which invalidates induction:

Turkey is fed everyday for several weeks at 9am. Is it safe for turkey to reason inductively and assume that it will be fed at 9am everyday during the month of December? Induction is sometimes justified by the fact that it is used in everyday practice and has worked well in Science so far. This is a circular argument. Nevertheless, inductive inference makes the conclusion highly probable. General scientific laws invariably go beyond the finite, limited amount of observable evidence that is available to support them. Therefore, scientific laws can never be proven inductively from sufficient evidence, but have to include deductive reasoning.

In good inductive arguments:

- the number of individual observations must be as large as possible;
- observations should be repeated under a wide variety of conditions;
- no observation should be in conflict with the derived laws.

4. Hypothetic-deductive method

Science cannot be based on the collection of facts alone or on logical reasoning alone. This is why a more complex procedure (scientific method) is necessary to discover new truths. This method integrates inductive and deductive reasoning. The inclusion of induction into the scientific reasoning process ensures that scientific truths are not always final (ultimate). The notion that Science and its results represent the ultimate Truth is called "scientism" and is an example of a misconception about Science (see discussion in Chapter 1).

The author of this book does not endorse "scientism", but this does not suggest that many opposing arguments about particular unresolved issues are all equally valid or true. The problem is that sometimes we cannot decide on the basis of available information which argument or theory is correct.

The following diagram outlines hypothetic-deductive method which combines induction and deduction described earlier.

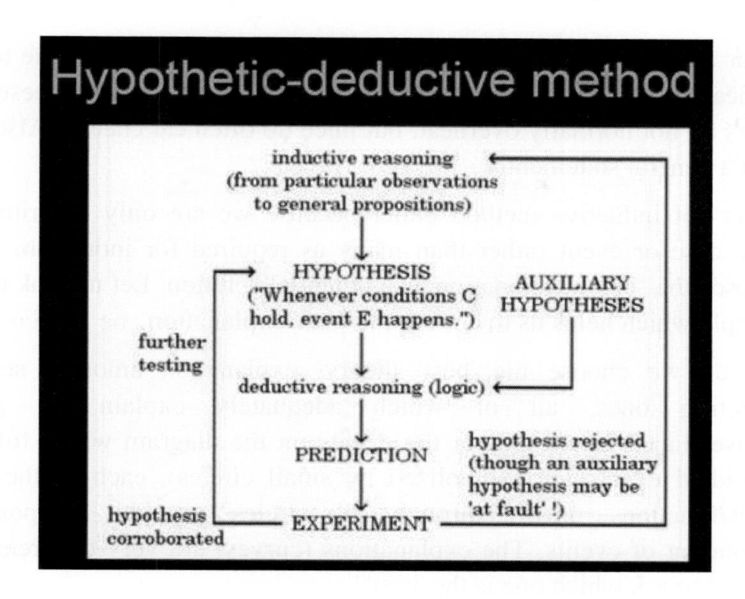

5. Inference-to-the-best-explanation (IBE) and Ockham's razor

IBE method is often used in Science because sometimes we do not know the appropriate general principle to start deductive reasoning from. On the other hand, the induction method has its own deficiencies as discussed earlier.

Consider the problem:

(1) The cheese in the larder has disappeared, apart from a few crumbs left over	(True)
(2) Scratching noises were heard from the larder last night	(True)
(3) The cheese was eaten by a mouse	(True?)

We can solve this problem by using IBE. IBE is not a deductive reasoning because (1) and (2) do not necessarily entail (3). Why? A clever maid could have stolen the cheese and left the crumbs to make it appear as if the mice did it. Also, scratching noises could be due to the overheating boiler. However, the maids do not usually steal cheese, the boilers do not normally overheat, but mice do often eat cheese! Also, (1) is not a general statement.

IBE is not inductive method either because we are only describing a single case or event rather than many as required for induction. IBE proposes the 'best' explanation as a potential solution. Let us look at the principle which helps us to choose the 'best' explanation.

How do we choose the 'best' theory/ explanation amongst several competing ones, all of which adequately explain **all** given premises/facts? If we look at the events on the diagram which follows (individual events are symbolized by small circles), each of the four curves/functions drawn through all circles provides a possible explanation of events. The explanations (curves) are very different and we need to ask which one is the 'best'?

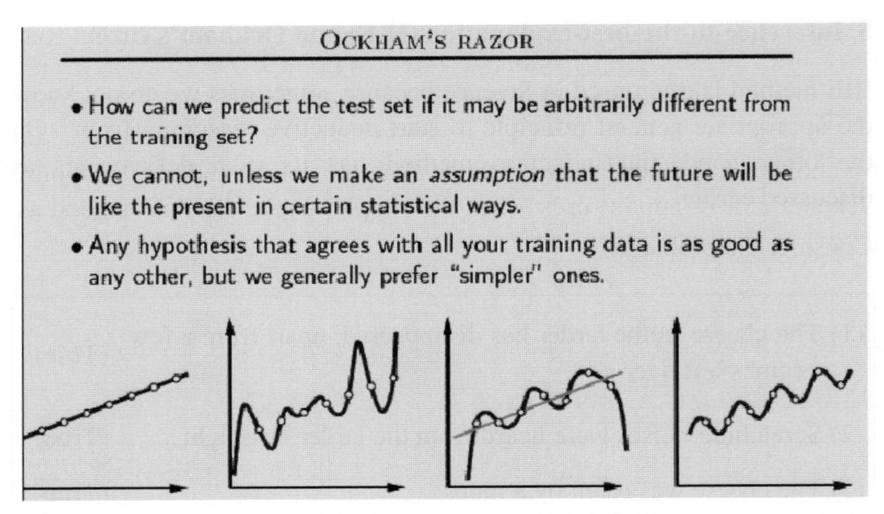

source: http://www.cs.ubc.ca/~murphyk/

Let us look at the principle which allows us to choose the 'best' explanation.

The principle of parsimony (Ockham's razor) states that:

Causes should not be multiplied beyond necessity; the simplest explanation is the best.

In the missing cheese example, a single cause (the mouse) is the simplest explanation because it explains both pieces of evidence/data (missing cheese and noises). A maid/boiler explanation postulates two causes (maid and boiler) in order to explain the data.

Darwin's Theory of Evolution explains a very wide range of observations, not just the anatomical similarities between species. Each observation could have been explained separately in a different way, but Darwin's theory made all the explanations in one go; it complies with the principle of Ockham's razor.

Ockham's razor principle [10] does not imply that the inference is necessarily true as is the case in deductive logic. Nevertheless, Ockham's razor is a very useful rule-of-thumb even though it cannot guarantee truth as a rigorous theorem can! Ockham's razor is a conservative principle which removes fanciful, complicated constructions. It curtails our psychological propensity to make elaborate stories and explanations. Ockham's razor applies **only** when the simpler explanation is as good as the more complicated one. One does not choose a theory or explanation for simplicity's sake alone! The selected theory must explain **all** the known facts **AND** at the same time be the **simplest** amongst competing theories or explanations (see the citation below).

The aim of science is, on the one hand a comprehension, as complete as possible of the connection between sense experiences in their totality and on the other hand, the accomplishment of this aim by the use of a minimum of primary concepts and relations.

A. Einstein

Why should we prefer the **simplest** amongst otherwise equally satisfactory models or theories?

- A simpler model/theory provides clearer and more comprehensible description.
- A simpler model/theory is more likely to be vulnerable to future falsification, because it has fewer adjustable parameters. Adjustable parameters can make any theory fit any set of data!
- It is known from mathematics that any function in which the number of adjustable parameters is equal to or greater than the number of observations, can be made to pass exactly through all points of the graph (i.e. predict exactly all the measurements *a posteriori*). The number of degrees of freedom (i.e. difference between the numbers of independent experimental observations and the number of adjustable parameters in the model) should therefore be large. This requirement favours the use of small number of adjustable parameters i.e. a simpler model. This is because the number of degrees of freedom reflects the number of points whose position is predicted without any constraints forcing the prediction to be correct.
- Relaxing the constraints embedded in Ockham's razor principle introduces the infinite number of possible hypotheses which will fit the available data.

6. Reasoning by analogy:

Example: Process 1 has similarities with Process 2, hence Theory 1 → similar to Theory 2

The most famous example of analogous reasoning was used as a starting point for Theory of Evolution. The theory of population pressure which describes selection/control in human society (proposed by Malthus) was used by Darwin to postulate theory of natural selection as the driving force behind the evolution of biological species (see Figure 7). Malthus has argued that human population growth outstrips the growth of food and other resources. This leads inevitably to competition and conflict in which stronger individuals prevail over weaker ones. Darwin's own words describe the event:

> *In October 1838 ... I happened to read for amusement "Malthus on Population", and being well prepared to appreciate the struggle for existence which everywhere goes on from long-continued observation of the habits of animals and plants, it at once struck me that under these circumstances favorable variations would tend to be preserved, and unfavorable ones to be destroyed. The result of this would be the formation of new species. Here then I had at last got a theory by which to work.* C. Darwin

Barlow, Nora, ed., The Autobiography of Charles Darwin *1809–1882. With the original omissions restored. Edited and with appendix and notes by his granddaughter Nora Barlow*, London: Collins, 1958

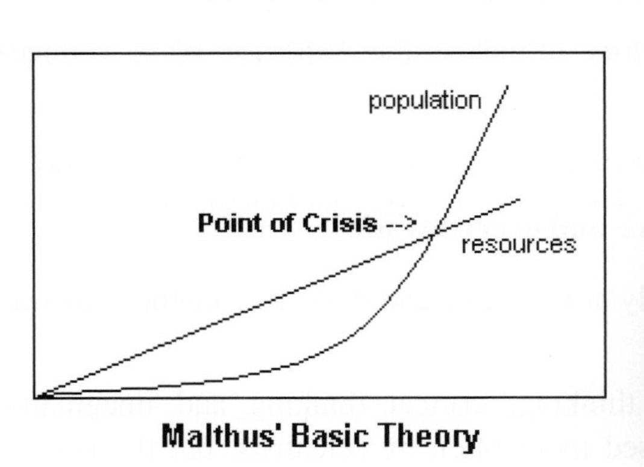

source: http://www.fordham.edu/halsall/mod/lect/malthus.gif/
http://www.nndb.com/people/250/000024178/malthus.jpg

Fig. 7.

Analogical reasoning should be exercised very carefully because its validity depends on the degree of similarity between the two processes/objects. The notion of "similarity" is often an imprecise, qualitative criterion! For example, in the Chinese tradition it is held that because rhinoceros is a big and strong animal, its ground horn can give virility and health to men. This is an example of analogical reasoning which is based on superficial likeness. The result of this reasoning was the near-extinction of the poor creature! While Darwin used analogical reasoning to steer him in the right direction, he did not rely on analogy alone. He used other types of reasoning to develop his Theory of Evolution.

Reflection:

Older generation often complains about the younger generation as being 'worse' than themselves. For example, you'll have often heard the phrase: 'When I was young ....'

Can this complaint be justified by using scientific arguments?

Tip: consider which factors influence the behavior of a younger generation.

Critical thinking in Science and everyday life

Now let's look more closely at inductive and deductive methods as used in practice.

There are two types of thinking: critical thinking and imaginative thinking. The former is used more often by scientists, but the latter is also important for developing new hypotheses and research ideas. Critical thinking is a careful deliberation on whether a certain logical argument/statement should be accepted, rejected or if the judgment on the claim should be suspended. Critical thinking includes estimating the degree of confidence with which the claim is accepted or rejected. Useful strategies for developing critical thinking are:

- **affective (attitudes and behaviour)** – examples of affective strategies include learning to think independently, exercising fair-mindedness, developing confidence in rational discourse.
- **cognitive (skills and process)** – examples include refining generalisations and avoiding oversimplifications, comparing analogous situations, clarifying issues, conclusions, beliefs, evaluating the credibility of information sources, recognising contradictions, developing ability to recognise errors and fallacies in arguments.

The validity of scientific knowledge depends on correct logical reasoning **AND** the quality of evidence. If either of the two is flawed, the conclusion may be wrong. Only if the logical argument (reasoning) **and** premises (evidence) are both true will the scientific conclusion necessarily be valid and true. Critical thinking thus involves consideration of both the soundness of logic and the accuracy of evidence.

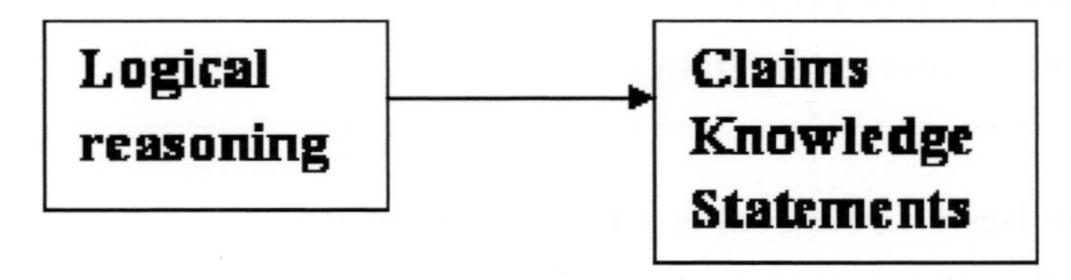

The Table below outlines possible relationships between evidence (premises) and logical conclusions.

Premises	Reasoning/logic	Conclusion
True	True	True
False	False	True/False
True	False	True/False
False	True	True/False

In examinations, students sometimes use erroneous logic and incorrect data and still arrive at the correct answer, although this is not very probable! This is why they are asked to explain their answers! Let's look at some examples which show how the correct conclusion (answer) can be arrived at even when the reasoning is incorrect.

a) Premises = F; Reasoning = T; Conclusion = T

All fish lay eggs (false)
Duck-billed platypus is a fish (false)
Duck-billed platypus lays eggs (true)

b) Premises = F; Reasoning = T; Conclusion = F

All cows are reptiles (false)
All reptiles can fly (false)
Therefore cows can fly (false)

c) Premises = T; Reasoning = F; Conclusion = T

Some art galleries don't charge an entrance fee (true)
London's National Gallery is an art gallery (true)
London's National Gallery doesn't charge an entry fee (true)

d) Premises = T; Reasoning = F; Conclusion = F

All witches keep black cats (true)
My neighbor keeps a black cat (true)
My neighbor must be a witch (false)

Logical arguments (deduction method)

Logical arguments are not personal conflicts/disputes. They represent a set of reasons and evidence in support of a conclusion or in an attempt to discover the truth.

Examples of correct logical arguments [9]

Correct logical arguments may be developed in a number of ways:

1. Modus Ponens

|If p then q | p | therefore q |

(also called 'affirming the antecedent'; the antecedent being ‘p’)

Affirming the antecedent

Person who support’s country’s war effort is a patriot	T
I am supporting the war effort	T
I am a patriot	**T**

2. Modus Tollens

|If p then q| not q | therefore not p|

(also called 'denying the consequent'; the consequent being ‘q’)

Denying the consequent

Corporate executives make good salaries	T
I do not make good salary	T
I am not a corporate executive	**T**

3. Hypothetical Syllogism

|If p then q| if q then r | therefore if p then r|

If I study I shall pass the exams	T
If I pass exams I shall graduate	T
If I study I shall graduate	**T**

4. Disjunctive Syllogism

p or q| not q | therefore p |

I have either passed or failed the exam	T
I have not failed the exam	T
I have passed the exam	**T**

5. Constructive Dilemma

| p or q | If p then r | if q then s | therefore r or s |

I have either passed or failed the exam	T
If I passed I shall graduate	T
If I failed I shall find a job	T
I shall either graduate or find a job	**T**

Incorrect logical arguments (fallacies)

The deductive argument cannot tell us if the premises are true or not. It is impossible to draw a conclusion which is stronger than the premises on which the conclusion was based. In other words, the deduction is 'truth-conserving'. If the premises are true and if they are combined in a logically permissible way, only then can we guarantee that the conclusion is correct.

When premises do not adequately support the conclusion or are combined in a logically unacceptable way, we have a *fallacy*. Note that the fallacious argument does not imply that the conclusion is necessarily wrong, only that we cannot be sure whether it is correct. Many fallacies often go undetected. Fallacies are not truth-preserving i.e. even though the premises are true, the conclusion does not have to be true.

Logical fallacies may arise due to:

- unacceptable premises
- reversal of conclusion and premises
- inadequate/insufficient premises
- irrelevant premises

1. Unacceptable premises

a. Begging the question / circular reasoning:

The conclusion is already contained in the premises. See examples of such reasoning below:

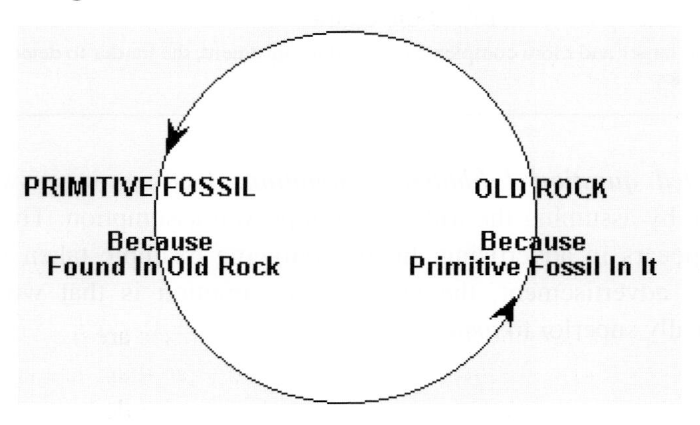

How do I know that the fossil is primitive? Because it was found in the piece of old rock. How do I know that the rock is old? Because it contains primitive fossils!

Circular Reasoning

A frequent error in arguments is to use the conclusion as one of the premises. This often 'begs the question' or involves circular reasoning, an argument which gives no support to the conclusion

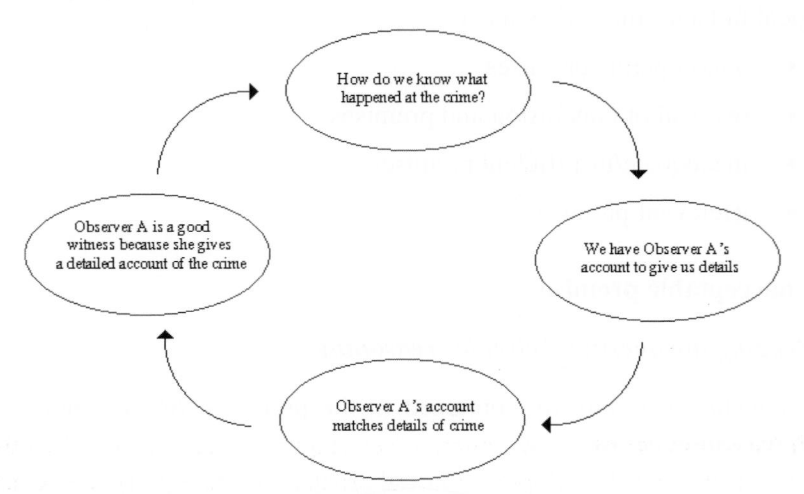

The larger and more complex a circle of an agrument, the harder to detect the fallacy

b. Loaded question / hidden assumption: Getting an answer to a question by assuming the truth of an unproven assumption. This fallacy often appears in advertising. In the following example taken from the cigarette advertisement, the unproven assumption is that women are biologically superior to men.

source: http://farm1.static.flickr.com/101/280853714_156f617ac4.jpg

***c. False dilemma*:** Only two alternatives (premises) are proposed as the positions that can be adopted. One alternative is inherently weak and easily discarded, which leaves the remaining one as true by default. This is 'we have no choice' type of argument. The problem with this reasoning is that there are often more than two possible alternatives, situations or decisions (premises). This fallacy is often used by lobbyists, politicians and journalists. Examples of this fallacy:

1. If you accidentally expose your film, then your photographs won't come out. (premise)
2. Your photographs haven't come out. (premise)
3. You must have accidentally exposed your film. (conclusion)

"If you are not for us then you must be against us!" This is a colloquial, shortened version of the logical argument which can be unpacked as follows:

1. You are either with us or against us on this issue. (premise)
2. You not with us. (premise)
3. Therefore you must be against us on this issue. (conclusion)

"We can have either good/secure jobs or livable environment!" (this is often heard argument about climate change)

2. Reversal of conclusion and premises

Denying the antecedent

If p then q| not p | therefore not q| (also called 'denying the antecedent')

Person who supports his country's war effort is a patriot	T
I am not supporting the war effort	T
I am not a patriot	F

The statement by Goering shows how the politicians can use this wrong argument.

"Naturally, the common people don't want war, but after all, it is the leaders of a country who determine the policy, and it is always a simple matter to drag people along whether it is a democracy, or a fascist dictatorship, or a parliament, or a communist dictatorship. Voice or no voice, the people can always be brought to the bidding of the leaders. This is easy. All you have to do is to tell them they are being attacked, and denounce the pacifists for lack of patriotism and exposing the country to danger. It works the same in every country."
Hermann Goering, Hitler's Reich-Marshall
at the Nuremberg Trials after WWII

source: http://www.myspace.com/berryentertaining

Affirming the consequent

|If p then q| q | therefore p|

For example:

Corporate executives make good salaries	T
I make a good salary	T
I am a corporate executive	F

3. Inadequate/insufficient premises

Inadequate premises include the following:

a. Premises give only partial support to the conclusion

b. Anecdotal evidence or hasty generalization (basing the general claim on a few atypical cases). For example:

Smoking is not harmful; my grandfather started smoking when he was a teenager and lived until the age of 90.

c. Omission or distortion of facts (creating a mystery by ignoring information which may provide alternative explanation). For example:

In Bermuda triangle - hundreds of boats and planes disappeared without a trace! However, in many cases wreckages were subsequently found. In any large oceanic area, near densely populated coastline (like Florida) there are likely to be disappearances due to accidents, storms, inexperienced pilots, sailors etc.

d. False cause (drawing a conclusion about the causal link which may not exist i.e. the apparent causality can be explained by coincidence or influence of other factors). For example:

<u>*A precedes B*</u>*: What happened after this must have happened because of this ("post hoc ergo propter hoc"). Just as I think of someone, the phone rings and that person is calling!*

Concomitant variation between A & B: *Over the last five years there has been great rise in the popularity of cellular phones and concomitantly the rise in the price of properties. The popularity of cellular phones caused the rise in property prices?*

e. Faulty analogy (justifying the novel explanation by emphasizing its similarity to the well-established explanation). For example:

Just as Moon influences the tides and the sunspot activity influences radio transmissions, so the positions of the planets have important influence on the formation of human personality.

f. Illicit ad hoc clauses or rescues (inventing auxiliary assumptions as means of explaining away a failed prediction, regardless of whether these assumptions are plausible or testable). For example:

If the rich are encouraged to grow richer then the poorest of the nation will benefit because the wealth generated by the rich will gradually trickle down to the poor. However, the effects of encouraging the rich to become richer were masked by the effects of recession. (Ad hoc clause)

4. Irrelevant premises

Irrelevant premises occur when the reasons given to support the claim are not related to the claim.

a. 'Straw man' argument: A caricature/misrepresentation of your opponent's views so that you can refute them easily; use of judgmental words such as corrupt or incompetent. This argument is often used by ideologues, politicians and journalists. For example:

Senator Jones says that we should not fund the attack submarine program. I disagree entirely. I can't understand why he wants to leave us defenseless like that.

Prof. Jones: *"The University has just cut our yearly budget by $10,000."*

Prof. Smith: *"What are we going to do?"*

Prof. Brown: *"I think we should eliminate one of the teaching assistant positions. That would take care of it."*

Prof. Jones: *"We could reduce our scheduled pay rises instead."*

Prof. Brown: *"I can't understand why you want to bleed us dry like that, Jones."*

b. 'Ad hominem' argument/genetic fallacy: *Shifting a debate from the point in question to some non-relevant aspect of the person making the argument. Again, this is a rhetorical trick used by ideologues, politicians and journalists. For example:*

We should not take seriously the findings of a medical scientist X who has researched the beneficial effects of jogging on cardiovascular system, because X is overweight and cannot run for more than a hundred meters.

Prof. Smith says to Prof. White, "You are much too hard on your students," and Prof. White replies, "But certainly you are not the one to say so. Just last week I heard several of your students complaining."

A prosecutor asks the judge not to admit the testimony of a burglar because burglars are not trustworthy.

c. Appeal to ignorance: Lack of known evidence against the hypothesis or theory is taken as a proof that theory is true. For example:

Since scientists cannot prove that global warming will occur, it probably won't.

No one has ever proved that silicon breast implants are unsafe. Therefore the silicon implants must be safe.

d. Fallacies of division/composition ("undistributed middle term"): The assumption that what is true for the whole is also true for the parts, or what is true of the parts is also true of the whole. For example:

Fallacy of composition

All cats are cute	T
All dogs are cute	T
All cats are dogs	F

(The middle term, 'cute', is not distributed between premises 1 and 2.) A proof that this is a fallacy can be made by the following Venn diagram:

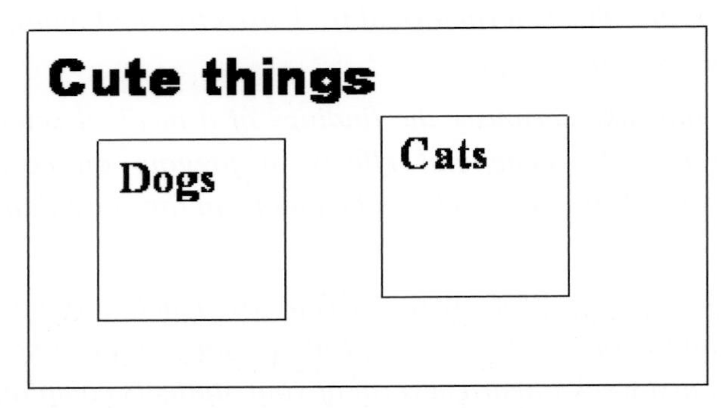

Another example of the same fallacy is given below (the undistributed middle term is 'number').

e. Equivocation: Two different meanings of the same word are used, but without a distinction being made. For example:

Organic foods are healthy, so any food with organic compounds in it must be healthy.

f. Appeal to authority, bandwagon fallacy, tradition fallacy: The past practice or authority in one area may not be used to support an idea in another area. Examples of the three fallacies follow:

This top athlete says eat breakfast cereal. (Athletes are not nutritionists).

But officer, I don't deserve a ticket; everyone goes at this speed. If I went any slower, I wouldn't be going with the stream of traffic. (bandwagon)

Astrology has been popular for thousands of years so it must be true. (tradition)

g. Gambler's fallacy is due to misunderstanding of the law of averages. Random events are independent of each other. For example:

The city of Miami did not have a major hurricane in 50 years. A hurricane is therefore due next year.

h. Appeal to emotion or fear ('slippery slope' argument): The proposition is assumed to lead inevitably to some terrible consequences (domino-effect). The terrible results are then argued against instead of the original proposition. This is used by ideologues, politicians and journalists. However, 'slippery slopes' can have different levels of steepness and we can decide how far down the slope we wish to go. For example:

"If you do not vote for us the opposition will get into power, destroy the economy and the country!"

"Euthanasia should not be legalized because it is the first step down the slippery slope leading to murder and genocide. Nazi techniques of mass extermination in concentration camps were first piloted as the euthanasia project to remove mentally and physically retarded from Society."

How to evaluate evidence or claims

When evaluating evidence in support of a scientific claim, we should consider:

- **Does evidence have contradictions?** Are there logical contradictions or contradictions with the well established core knowledge? For example, any design which proposes the creation of energy out of nothing can be rejected out of hand because it contradicts laws of thermodynamics.
- **Does the quality and quantity of data justify claims put forward?** Scrutinise quantitative data, the evidence presented as series of numbers, statistics. Check evidence for factual accuracy, precision of numerical data, sampling bias (e.g. has evidence for the efficacy of a drug been obtained by the survey of cured patients only etc). Do numbers conform to the expected order-of-magnitude estimate? Is evidence experimental or anecdotal?
- *For example: 25000 tons of a certain metal are mined per year. How important is it to recycle it? The evidence is insufficient to decide about recycling! We must know the estimated reserves of this metal in Earth.*
- **Is evidence relevant to the claim?**
- **Are there hidden assumptions related to evidence?** The premises may not be stated clearly and may contain hidden assumptions which in turn need to be assessed for validity. For example:

Biological organisms can be classified into plant and animals species. This argument carries the hidden assumption that all organisms must fit into one or the other category.

My friend is clever so he does well in the exams. Hidden assumption is that clever people do well in the exams.

- **Are the sources of evidence impartial? Have they been cross-checked? Are they informed? Have they been cited?** If the source of evidence is not given ("Many scientists agree that …") or if only a single source is given, the evidence may be suspect. Does the source know (or could know) the events/facts? For example, the biographers who describe lives of famous people often talk about how the person felt/thought at a particular moment. Yet only the subject of biography knows that!

N.B. Logical fallacy does not depend on the truth or falsehood of individual premises. Fallacy arises when the premises are combined in an inappropriate way and therefore do not support the conclusion!

Reflection:

Put the statements A-C in correct logical sequence.

A. Saturated esters are a type of chemical compound which produces the peak in IR spectrum at 1740 cm^{-1}.

B. The studied compound produces a peak at 1740 cm^{-1}

C. The studied compound is a saturated ester

Tip: consider if the logical argument demands that only esters produce a peak at 1740 cm^{-1} ?

Experimental Design and Scientific Evidence [11]

Why is the understanding of scientific evidence important? It is because it helps us to make the best possible decision, under given conditions. For example, which method of birth control should we use, whether to quit drinking or not?

How can evidence be used? We use evidence to search for **relationships** between objects, phenomena. Knowledge of causal links also helps us to make **predictions** and thus adapt our behavior to the circumstances. For example, the causal link between the mass of an object suspended from the rubber band and the extent of band stretching tells us how far the band shall stretch for a given suspended mass. This leads to a decision: how heavy an object may I attach to the band without it snapping?

We can use evidence to establish **difference** between groups of objects, or events. For example, we may examine the difference in the recovery rate of cancer patients treated in hospital A and hospital B? Evidence will then help us to make a decision on whether to choose hospital A or B for medical treatment.

We can use evidence to establish **change**. For example: How do nutritional contents of certain foods vary with length of storage? On the basis of evidence gathered we can decide which type of food is suitable for storage at home.

The concept map below shows criteria regarding the quality of evidence which must be satisfied before we can draw conclusions from that evidence. It is important to note that evidence must be both, reliable and valid in order to be useful.

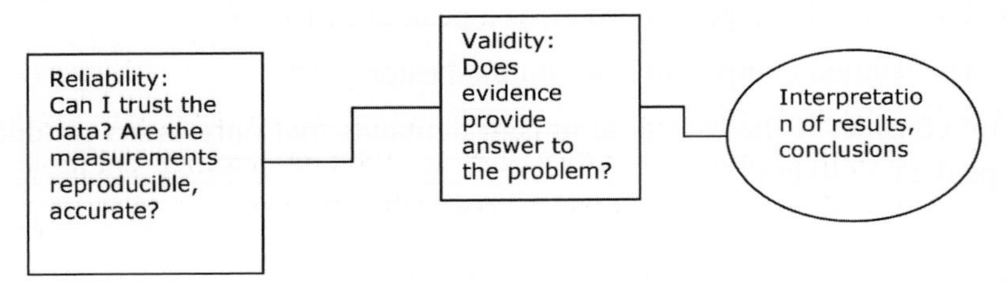

To answer questions and solve problems in Science, we perform investigations and experiments. The design of such experiments is crucial for determining whether our evidence and conclusions will be valid.

Phenomenon being investigated usually contains a number of factors influencing it. Such factors are called variables. To every variable a value can be attached. Variables can be classified in different ways. When classified according to their relationship to each other we have three types of variables in experimental design:

- independent variables (they can vary independently of the rest)

- dependent variables (they vary in response to independent variables)
- control variables (are kept constant during the experiment)

The concept map which follows ('circle of variables') shows how a simple experiment may be designed and how variables relate to each other.

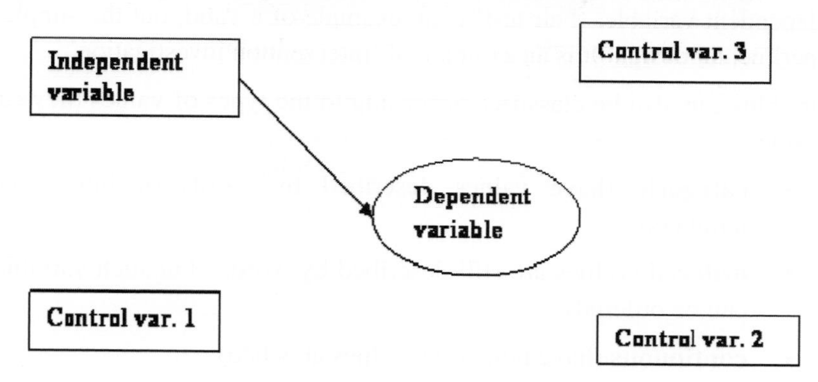

Study of the phenomenon involves answering specific questions about it. The questions regarding the phenomenon are formally expressed as links between two variables (one independent, one dependent) in the circle of variables. All key variables in the experiment must be taken into account for the design to be satisfactory. Let us look at another example of experimental design which aims to answer the question:

How does the shape of the kettle affect the time required to boil water?

Dependent variable DV (exhibits the effect) = time required to reach boiling point
Independent variable IV (causes the effect) = shape of the kettle
Control variables CV = volume of water, size of the heater, kettle material, starting temperature of water, electrical resistance of the heater.

We can choose a different variable as independent and link it with the same dependent variable. The research question would then change and become:

How does the electrical resistance affect the boiling time?

The investigated problem determines which amongst the key variables is independent and which is dependent. All other variables then become control variables (CV). CV must be kept constant during investigation/experiment or it must change uniformly for each value of the independent variable. 'Fair test' is an example of a valid, but the simplest experimental design. It is an example of 'intervention investigation'.

Variables can also be classified according to the types of values they can assume:

- **categoric** (have values described by words or labels, not numbers)
- **ordered** (values are still described by words, but such variables can be ordered)
- **continuous** (have numerical values attached)
- **derived** (from other variables e.g. speed km/h)

Examples:

Variable (type)	**Value**
Kettle material (categoric)	copper, steel
Kettle size (ordered)	small
Kettle volume (continuous)	100 ml
Kettle heating rate (derived)	deg/second

Variable types determine whether nominal, ordinal or interval data are obtained in the experiment. There are nine possible types of investigations depending on possible combinations of types of variables. The investigation types cover the range from:

ordered-ordered .... continuous-continuous.

The experimental design where both variables are of continuous type provides the strongest evidence/most useful information. The same holds for control variables. However, sometimes investigations with all continuous variables are not possible or necessary. For example, the investigation below falls into the continuous-ordered type.

Anti-depressant drug/mg	Depression state
5	Deep
10	Mild
20	Negligible

Intervention investigations:

1. Lab experiment (IV is changed in a controlled way by the experimenter)
Investigation: How does the rate of reaction depend on temperature? How much water does the tomato plant need for growth?

2. Randomized control trials (RCT) (IV is changed by experimenter but there are many CV present which cannot all be kept constant)
Investigation: Is the new drug effective against breast cancer? There are many possible CV here: age, sex, medical condition, fitness, diet, occupation, intake of other drugs, stage of cancerous disease etc.

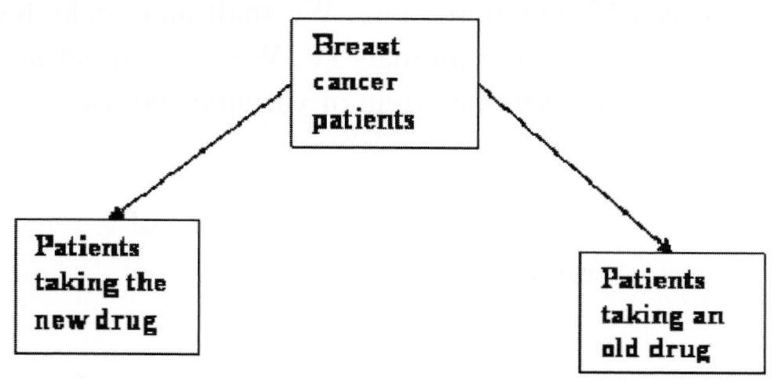

The large number of patients and their random selection will ensure that CVs in the two groups (experimental and control group) are evenly matched and thus the relationship between IV and DV can be determined. Such experimental design also minimizes experimental bias (sampling bias, psychological bias, subject bias etc.). In **blind studies** and **double-blind studies** a 'placebo' is used to remove bias in the experimental results.

Non-intervention investigations

1. Field study (no variable is changed in a controlled way by the investigator)

For example: *How do elephants affect vegetation in the national park?*

The investigator collects data on variables which are of interest or are relevant. These data are used retrospectively and there is no change in the natural situation (no intervention). IV and CV may be interchanged and we can thus create new questions as we proceed with the investigation.

What factors affect the behavior of a particular gazelle species in a particular national park?

We may start by examining temperature effect first i.e. by taking temperature as IV, but find subsequently that the presence of other animals (which is a CV) matters more. We shall then make temperature CV and the presence of other animals IV. We are generating questions from the available data! See the circle of variables below.

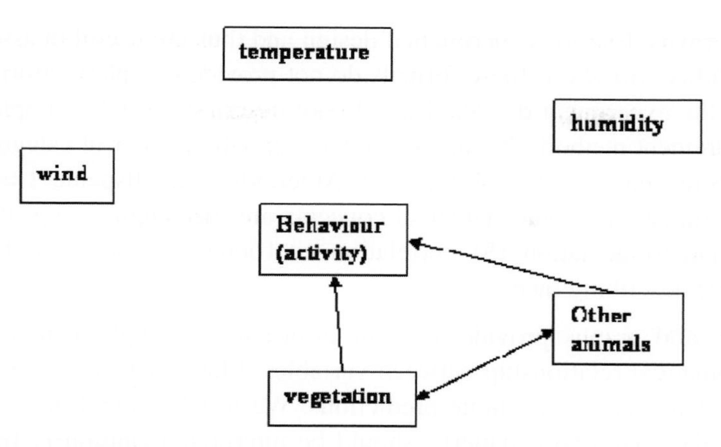

Scientific evidence (data) can be presented in two main formats: tables and graphs. The best format to use depends on the nature of data and variables. The presentation formats describe structure of the experiment (IV, DV but not CV), units, number, range and interval of experimental readings. These formats have two purposes:

- to help organise the investigations, record measurements; and
- to display and present data after the measurement, to optimise range and interval of measurements.

IV	*DV*	*Type of data*	*Best format*
categoric	continuous	Nominal	Bar chart
ordered	continuous	Ordinal	Bar chart
continuous	continuous	Interval	Line graph, histogram
categoric	categoric	Nominal	Table
ordered	categoric	Nominal	Table
ordered	ordered	Nominal	Table

The formats describe experimental design and thus are useful in assessing its validity. However, these formats do not provide complete information about the experiment design. They do not describe e.g. CV, sample size, measurement method. (Nominal data are data where order of categories is immaterial e.g. 1=other, 4=black, 3=Amerindian, 2=Hispanic 1=white). For nominal data some statistical concepts are meaningless; e.g. there is no standard deviation (S) correlation coefficient (r) or mean for the variable describing race.

Tables and graphs provide a lot of information. Graphs can reveal a mathematical relationship between variables. Mathematical relationships are best for making accurate predictions. We usually search for patterns in the data. However, patterns should be interpreted cautiously because data may have limitations; alternative interpretations may be possible. For example, the type of data obtained may limit the extent of predictions or generalizations about the phenomena. The relationship between variables is easiest to detect if at least one variable is continuous. **However, correlation between variables does not necessarily imply causal link!** Important guidelines for understanding data include considerations:

- Is there a pattern revealed by the graph or table?
- What the data do not tell us?
- Can we predict or generalise on the basis of data?

Let us look at the example of the experiment where the relationship between car speed and stopping distance was investigated (below)

Speed (m/s) IV, continuous	**Stopping distance (m) DV, continuous**
10	15
17	30
20	42
23	57
30	90

Since IV and DV are both continuous, the data can be readily interpreted to suggest that there is relationship between speed and stopping distance. The graph (below) gives us an even better insight into this relationship.

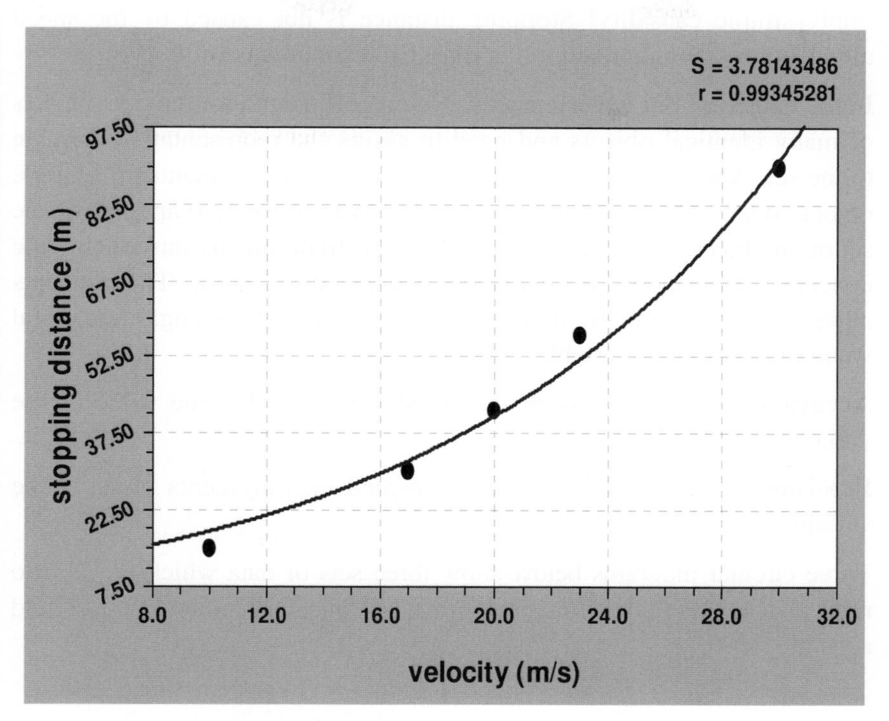

What does the data tell us? It tells us that as the speed of the vehicle increases, so does the stopping distance. Also, the stopping distance increases with speed more rapidly above the threshold velocity of 17 m/s.

What doesn't the data tell us that may invalidate our interpretation? The data doesn't provide information about the type and number of cars used in the study, how accurately the stopping distances were measured, the road surface condition, whether the test was done only once at each speed, what happens at velocities below 10 m/s and above 30 m/s and so on.

Does the data suggest association/correlation, difference or change between the two variables? The data suggest a direct association/link between speed and stopping distance. However, this does not necessarily imply simple causality! Stopping distance is not caused by the speed alone, but by the momentum (car mass) and friction as well.

In the course of our experiments we often perform measurements on sets of many identical objects and need to assign the representative variable to the set. Alternatively we perform a series of measurements on a single object. What value can we assign to a specific property (variable) of the set or an object? Statistical analysis helps us to do this by introducing the concepts of accuracy, precision and sample distribution. The concepts allow us to reduce many individual measurements to a single numerical value of a particular variable.

Accuracy – the closeness of measured values to the true value of the variable

Precision – the variation between repeated measurements of the same variable

Three circular diagrams below show three sets of data which are (left to right): inaccurate and imprecise, precise but inaccurate, precise and accurate.

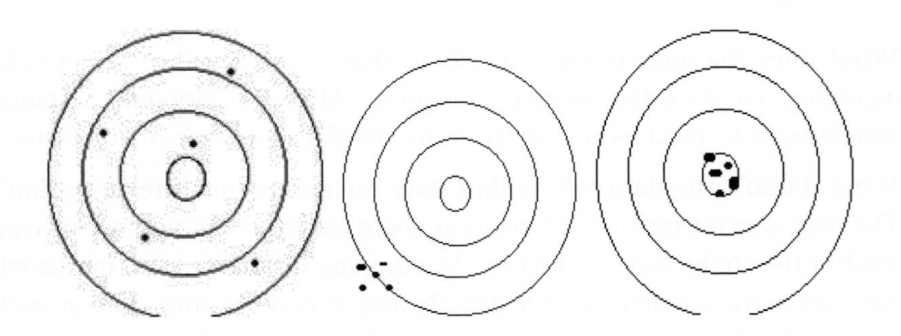

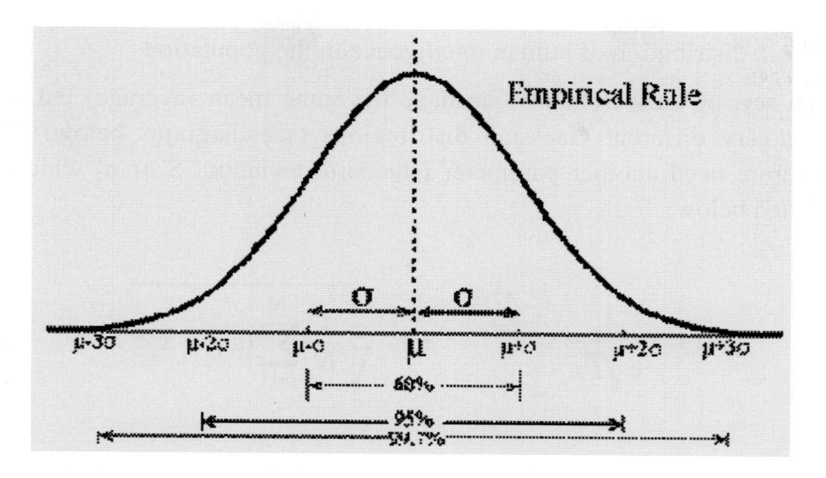

The Gaussian distribution (shown above) of measured values applies in situations where each measurement is affected by a large number of small, independent, random errors. These errors are of the same order of magnitude and comprise equal number of positive deviations/errors (measured value is larger than the 'true' value) and negative deviations/errors (measured value is smaller than the 'true' value). Because of its form this curve is also called the 'bell curve'.

When measuring a physical variable, one tries to eliminate systematic errors, so that only random errors have to be taken into account. In such case, the measured values will be distributed around the average value of the variable, as in the Gaussian curve. It can be proved that the average of measured values is the 'best' or 'true' value, if Gaussian distribution holds for measured data. The 'best value' is defined as the value, for which the probability of occurring during subsequent measurements is greatest.

Some real life examples following Gaussian distribution are:

- distribution of the body heights of individuals (of a given sex)
- distribution of weights of machine packed boxes (containing washing powder)

- distribution of human intelligence in the population

Two sets of measurements can have the same mean (average) but still have very different Gaussian distributions (see diagrams below). We therefore need another parameter (standard deviation; S or σ) which is defined below:

$$y = \frac{1}{\sigma\sqrt{2\pi}} e^{-\frac{x^2}{2\sigma^2}} \qquad \sigma \equiv \sqrt{\frac{1}{N}\sum_{i=1}^{N}(x_i - x)^2}\,.$$

The points of inflection on the curve are situated at x = ± s. For this distribution two out of three measurements differ by less than S from the maximum/average value. Approximately one in twenty measurements differs by more than 2S from the average.

The three sets of data mentioned earlier are now shown as respective distributions.

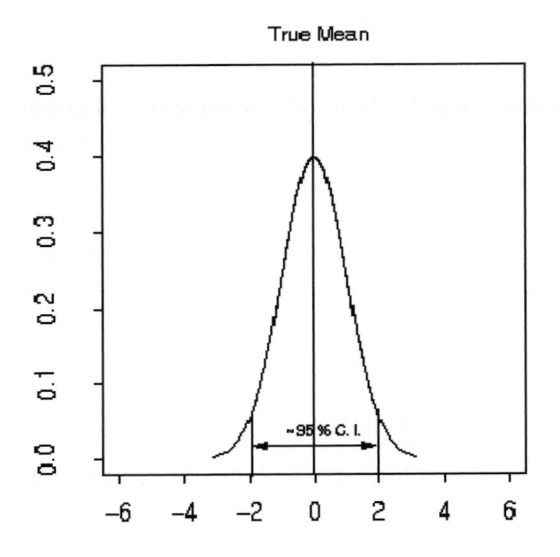

Population:
$\mu = 0$
Sample:
$\bar{x} = 0 \quad s = 1$
Accurate and Precise

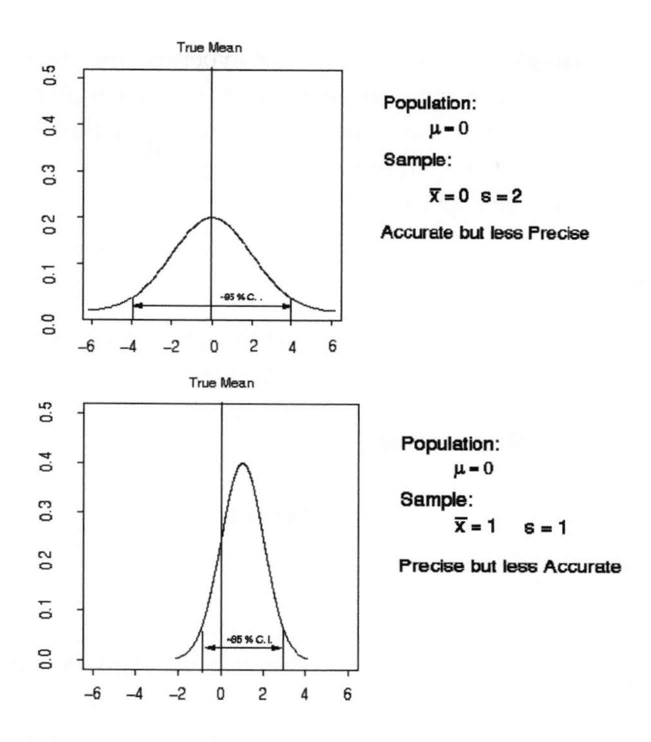

Measurement errors

- **Human error** (incorrect reading/use of instrument, incorrect choice of an instrument) is avoidable through e.g. staff training and is not subject to statistical analysis
- **Systematic error** (improper calibration, zeroing etc.) is difficult to discover except by reference to a standard set of data. This error is avoidable through calibration, but cannot be described by statistical analysis.
- **Random error** (described as uncertainty in measurements) is unavoidable and must be analysed statistically.

The errors/uncertainties can be associated with: measuring instrument, measuring process or the variable being measured (e.g. blood pressure increases when it is being measured- "coat hypertension", blood pressure also varies daily). When reporting measurements, uncertainty should be calculated and reported. This helps us to decide whether the data are reliable enough for our purpose. "Reliable enough" depends on the context and task of the experiment.

Sampling

Experiments often cannot be performed on the whole population of objects so the representative sample of the population must be selected instead.

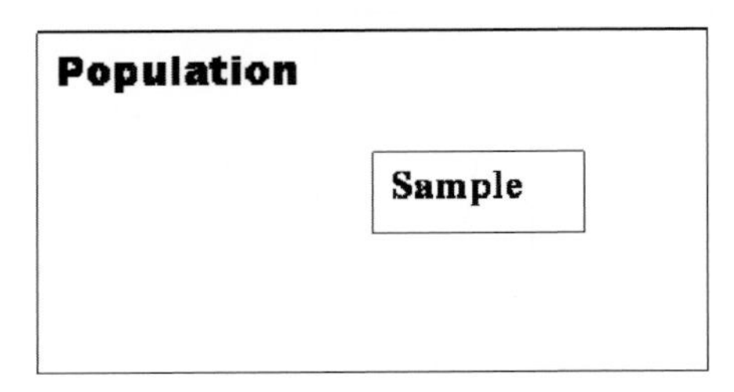

When describing data/measurements we use standard deviation S (σ) rather than mean (average) or range. This is because σ describes how are the data spread around the average, it quantifies the shape of data and allows prediction of probability of subsequent measurements of the same variable. The spread of measured values is due to "natural variation" of the sample and to measurement error. However, σ should be calculated only if 20 or more measurements of the same variable are available.

S describes only the distribution of measurements within the sample. What about the distribution in the whole population? If the specimens in the population are distributed normally (Gaussian distribution), we can assume that the distribution of mean values for different samples will also be normal. Then we can calculate standard error:

$$SE = \sigma / \sqrt{n}.$$

SE generalizes from the sample to the population! Our measured value should then be expressed as (average ± SE). The smaller SE, the more precise the measurements are. We can reduce SE by increasing the number of measurements (sample size) or the accuracy of measurements. If we are measuring a single variable the standard error can again be reduced by repeated measurements. In order to reach a conclusion from the data, sample must be sufficiently large. How large? That depends on how reliable or probable we wish our conclusion to be!

Sometimes we wish to determine whether there is a difference between two populations (e.g. A, B) or if there is a difference between measurements of the same population at different times. We then take large number of measurements and check whether the two intervals ± SE overlap. **If intervals do not overlap, we can say with 68% confidence that A and B are different populations.** When searching for patterns in the data we must also ensure that SE are sufficiently small to justify drawing conclusions about relationships.

Reflection:

Two chemists (A & B) used the same procedure to measure the content of element carbon in the sample by making 5 replicate measurements each. The average values obtained by the two were 51.2% and 50.9%, respectively. The standard deviations obtained by A was 3.7 and by B 4.5. The conclusion may be drawn that chemist A is the more consistent worker than B.

Which of the statements A)-D) best describes the flaw in the conclusion above?

A. There may be a systematic error in the measurement.

B. The difference between two standard deviations is not big enough for the two values to be significantly different from each other.

C. The standard deviation gives no information about accuracy.

D. the confidence limits of the mean are given by the standard error of the mean and not by the standard deviation.

Tip: consider the uncertainty intervals which apply to the two chemists and see if they overlap. Also consider the number of measurements performed by each chemist.

Pseudo-science (PS)

The importance and respectability of Science (S) in modern Society has lead to attempts to disguise scientifically invalid thinking as "Science". It is thus important to be aware of the distinction between Science & Pseudo-science (PS). PS is also a nice example to illustrate how Science should or should not be done.

Why is PS accepted? Possible reasons include:

- poor critical thinking ability which cannot separate real from unreal events
- inability/unwillingness to scrutinize sources of information or to interpret the evidence correctly. *People accept testimonials from patients who believe were cured by a certain medicine without trying to obtain information from the patients who were NOT cured. People have a natural tendency to evaluate propositions they agree with differently from those they dislike. Many people tend to believe unquestioningly the information provided by media, books etc., especially about things they want to believe in ("wishful thinking")*

Examples of PS are: *creationism, "perpetuum mobile" devices, graphology, homeopathy, paranormal phenomena, UFO-s,* Velikovsky's *"Worlds in Collision".*

The best antidote to PS is the development of critical thinking ability.

Characteristics of PS:

- PS always claims to be Science
- S fully and rigorously tests new ideas, PS does not.

The distinction between S and PS cannot be made on the basis of the appeal of results/ideas produced by the two disciplines. It matters not only what you say, but also how sound is the method/evidence, which you used to arrive at your result/proposal.

- PS involves static or randomly changing ideas;
- S involves progress, accumulation of knowledge, increase in the volume of evidence and theories. PS proposes spectacular claims and dogmatic ideas which do not change with time or change only randomly. This is because PS has no roots in the well-organized body of knowledge nor does it foster systematic comparison of theories with experiment. The PS theories are static because PS has no established mechanism by which to evaluate or reject ideas, so there is no reason to accept some PS theories and reject others. As a consequence, S is self-correcting while PS is not. Peer review system in scientific journals ensures correction of potential errors and purge of bogus results. However, Science journals are also a forum for criticism of previously reported scientific research. In PS journals (e.g. creationist journals) there is no criticism by one creationist of the work performed by another. PS uses vague mechanisms to acquire understanding, uses loosely connected thoughts and puts the burden of proof on the opponents.

Understanding in PS is based on incoherent premises which are not consistent with the observations. Ideas in PS may be so vague that they can explain everything (and hence have no value), or they may be based on weak analogies e.g. the link between sunspots & business cycles. Donnely in 1882 proposed the existence of a sunken continent Atlantis on the basis of some similarities between ancient civilizations of Egypt & South America: pyramid building, flood myths, embalming etc. SP reasoning involves logical errors e.g. proof-by-ignorance. 'If you cannot prove my claim wrong it must be correct'. Another example of logical fallacy which appears in arguments used by PS and Creationists is shown below:

Question to the Creationist: *How could Noah have possibly fitted dinosaurs in the Ark?*

Answer by the Creationist: *How was it possible for your mother to give birth to you?*

The fallacy is faulty analogy.

- S involves organized scepticism, while PS does not. PS regards sceptical attitudes as a sign of 'narrow-mindedness' of Science.

This lack of scepticism leads to PS accepting claims which are not based on solid evidence. Scepticism in Science may fosters conservativism (new ideas are not immediately accepted in order to ensure that bogus results are removed), but also opens the possibility for change. Scepticism in S is extended to old ideas as well as new; when contradictory results after the initial scrutiny accumulate, old ideas are replaced. Scepticism in PS is applied selectively, mostly to the ideas of PS opponents not to the ideas of PS supporters. The prospective members of 'Creation Research Society' have to sign a manifesto affirming the tenets of Creationism before joining the society!

- PS disregards well established and proven scientific results. Contradicting established results is considered by PS a justifying virtue for the new, exciting idea. In S new results/ideas have to be integrated into the existing body of knowledge and are subject to particular scrutiny if they contradict the established knowledge. Practitioners of PS work in isolation from broader intellectual community (scientists work within the scientific community) and thus have poor understanding of the topic of their study. PS communicate results directly to general public (without peer review) which similarly lacks deeper understanding of the topics. The self-consistency between all concepts and ideas is one of the greatest strengths of S and simultaneously the greatest weakness of PS.

Improper methods used by pseudo-science include:

- anachronistic thinking (using theories discarded long time ago)
- seeking mysteries (deliberate search for anomalies in Nature in the belief that the new theory which explains the anomaly can be accepted) like including UFOs, Yetis, the Bermuda triangle.
- use of ancient myths and literary interpretations of old texts as evidence (assuming that ancient myths are literally true and

devising hypotheses to explain them). For example, Velikovsky based his analysis of past geological and astronomical events on ancient manuscripts, legends, traditions. He thus 'explained' the parting of the Red Sea, manna from Heaven, the Egyptian plague and so on.

- casual approach to evidence: use of large amount of data irrespective of their quality and refusal to weed out dubious data. For example, Velikovsky claimed that the direction of the Earth's rotation changed in the past on the basis of many textual references in ancient cultures. Science objects to this approach because: How reliable are the quotations? Do they all refer to the same date? What was their context? No alternative explanations were considered by Velikovsky.
- reliance on spurious similarities as evidence, use of superficial similarities with the established scientific results
- explanations of events by scenario in disregard of the established laws/principles. For example, Velikovsky postulated that a large planet-sized object was ejected from Jupiter, became a comet, passed close to Earth causing plagues in Egypt. Earth then passed through the tail of the comet which was composed of hydrocarbons. The hydrocarbons fell on Earth as 'rains of fire' (petroleum?) and manna (carbohydrates?). His explanations of the events described in Bible use physical and not supernatural causes. However, his explanations also violate principles of classical mechanics (conservation of momentum) and do not propose alternative mechanisms by which these events might have taken place.
- refusal to revise ideas. Cranks and crackpots never admit that they are wrong; they use rhetorical arguments and never revise their position even after justified criticism.
- proposing unfalsifiable explanations or predictions. Proposing an explanation which is consistent with anything/everything that can occur; or making a prediction which is so broad or vague that it will be fulfilled no matter what happens.

An example of unfalsifiable prediction is the well known prophecy made to ancient king Croesus. King Cyrus the Great of Persia, in the process of extending his vast empire threatened the kingdom of Lydia. The Lydian king, Croesus, consulted the oracle of Delphi in Greece about whether to start the pre-emptive war with Persia or not. The oracle replied: *'If Croesus goes to war he will destroy a great empire.'* On the basis of this prophecy Croesus went out to meet the army of Cyrus and was utterly defeated. He destroyed the great empire, his own! Another example of unfalsifiable proposition is when Creationists claim that Universe was created by God via processes which no longer operate in the natural world e.g. worldwide flood.

When evaluating claims which purport to be scientific we must:

- ensure that the claim is stated clearly and specifically
- examine the evidence presented in support of the claim
- consider alternative hypotheses
- rate hypotheses according to their degree of adequacy

The adequate hypothesis must be testable, fruitful (theories which explain more than was originally intended), broad in scope (the more the theory explains the better it is), simple (Ockham's razor) and conservative (rationalize all the previous knowledge). Creation 'science' is a good example of pseudo-science and we shall describe it next.

Creation science (CS)

source: http://en.wikipedia.org/wiki/File:The_Creation_of_Adam.jpg

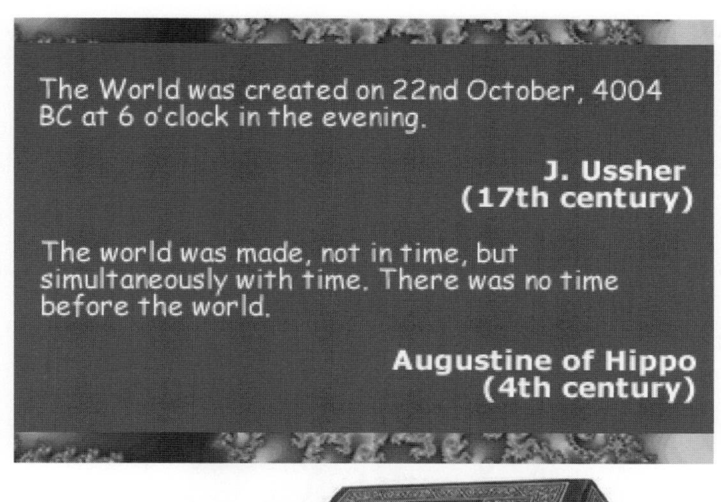

source: http://www.middle-east-online.com/pictures/big/_17686_koran-2-10-2006.jpg

Koran: [49.1] Say, 'You disbelieve in the One who created the earth in two days, and you set up idols to rank with Him, though He is the Lord of the universe.'

[41.12] Thus, He completed the seven universes in two days, and set up the laws for every universe. And He adorned the lowest universe with lamps, and placed guards around it. Such is the design of the Almighty, the Omnipresent.

Fig. 8. Michelangelo's painting and Koran

source: http://www.wpclipart.com/world_history/Darwin_ape.png.html

Fig. 9.

Figures 8, 9 and citations above illustrate links between Creationism and monotheistic religions and also the depth of controversy which CS has created. CS does not explicitly claim religious connections, but such connections are too obvious to deny. It is interesting to note that not all religious people interpret sacred texts literally as the quotation from Augustine of Hippo shows. This is an important point especially bearing in mind that Augustine lived in 4th century AD. Most Creationists however do adhere to literal interpretations of religious texts.

Tenets of Creationism

- Sudden creation of the Universe, energy and Life out of nothing.
- The insufficiency of mutation and natural selection in bringing about the diversity of existing species.

- Only changes within fixed limits of originally created kinds of plants and animals are possible.
- Separate ancestry for man and apes.
- Earth's geology is moulded by global catastrophic events (catastrophism), including worldwide Flood.
- Relatively recent inception of Earth and living species (several million years).

Creationism's objections to evolutionary theory (ET)

According to Creationism, Evolution is perceived as an attack on religion and morality. This is a non-scientific objection, because it is related to the question of values. The crux of the argument is not whether ET has moral implications or not, but whether it is consistent with what we discovered empirically about Nature. In fact ET has no moral or religious implications [12] as shall be discussed in Chapter 3! Ethics provides guidelines regarding human behavior which Science cannot give. Therefore, possible threats to Ethics suggest that some of our ethical norms may be based on principles divested from practical experience. This does not invalidate those ethical principles, but only explains why the conflict with Science occurs. Here are more examples of arguments used by CS:

> ***"For as a Man thinketh in his heart so [is] he"***
>
> *Proverbs 23:7*

This is fallacy called *false analogy* and a conflation of Ethics with scientific reasoning.

The cartoon in Figure 10 is a more complex example of *slippery slope argument* conflated with *hidden assumption* fallacy. It is trying to portray Christianity as morally superior to Atheism and rational doubt as a path leading to moral corruption. Scientific reasoning is simply a method which keeps us 'in tune' with the Reality, it has no moral implications *per se* and was instrumental in confining enormous practical benefits on

ILLUSTRATION FROM WILLIAM JENNINGS BRYAN'S SEVEN QUESTIONS IN DISPUTE, 1924.

source: http://www.freesundayschoollessons.org/historical-theology/baptists-influence-or-separate/

Fig. 10.

humanity. It is true that Science cannot answer several important questions about human life, but that does not make it a morally corrupting influence. The extrapolations from Science which are being made in the name of Science are a different matter, but such extrapolations do not belong to the realm of Science.

Interestingly, similar non-scientific objection, but this time against Religion, was raised by Dawkins [13a] who claims that one of the reasons why Religion should be stamped out is because it poses a threat to modern civilization. Dawkins's claims have been challenged by both atheist and religious thinkers [13b,c]. Both objections described above are missing the point. The issue at stake is not whether Evolution is a threat to morality or not, but whether it provides a consistent and verifiable rationalization of empirical data. Likewise, Dawkins argument about getting rid of Religion because it is dangerous is also inadmissible. Science itself can be considered dangerous if we recall the existing or potential threats originating from scientific discoveries e.g. weapons of mass destruction, cloning of humans etc. Yet, Dawkins has not proposed that we get rid of Science. The issue to be decided is whether Religion adds something new to humankind which is not provided by other elements of culture including Science. If Religion is useless, this would be sufficient reason not to take it seriously even if it were harmless!

Creationism is *not* Science [13d] but it *does* highlight an important problem. The problem is whether human values should or could be derived from understanding of the physical World and extrapolations of ET to this World or from other considerations. The suggestion that Ethics can be derived from scientific considerations is an example of 'naturalistic fallacy' which shall be discussed in Chapter 3. CS opposes *moral individualism* which suggests that individual human/animal beings should be treated not according to their group membership, but by considering individual's own characteristics (e.g. reproductive fitness, genetic health, physical strength and so on). As mentioned in Chapter 1, some modern philosophers like Singer have argued that a healthy, fit animal is more 'valuable' than a disabled human. Such comments together with similar ones by Dawkins (below) provide driving force for CS and create confusion by conflating ethical and scientific issues.

'We are survival machines-robot vehicles blindly programmed to preserve the selfish molecules known as genes. This is a truth which still fills me with astonishment.'

R Dawkins

Leaving moral conundrums aside, what rational arguments does CS put forward in support of its propositions? Its main arguments are outlined below.

The living organisms are too specialized (organized) to be a product of random processes ***...life is too complex to have emerged without divine intervention in the form of an 'intelligent designer'.*** (Intelligent Design theory)

This fundamental question has been debated for a long time and can be rephrased as: Do complex effects and structures **always** have complex causes? The answer is NO (as we shall soon see), but the problem remains unresolved because we still cannot tell when the complex effects shall be due to complex and when to simple causes. The description of 'simple' and 'complex' is also dependent to some extent on the world view which a person assumes to be true. When does simple become complex? What quantifier can be used to describe the transition from simple to complex?

There are sharp differences of opinion on this issue within and without scientific community. The same differences of opinion are closely related to the two World views: reductionism and holism which shall be discussed later.

a. Conservation of complexity

Must complex effects have complex causes? Our intuitive answer would be yes. We have direct experience of many instances where a sophisticated, complex technological devices, theories or works of art are indeed constructed or created by brilliant, sophisticated minds. Computers, watches, aircraft, theory of relativity, mathematics, art paintings are some of the many well known examples. The question is however whether this always happens or whether there are instances when no complex causes are at work.

source: http://www.wpclipart.com/world_history/ape_man_evolution.png.html

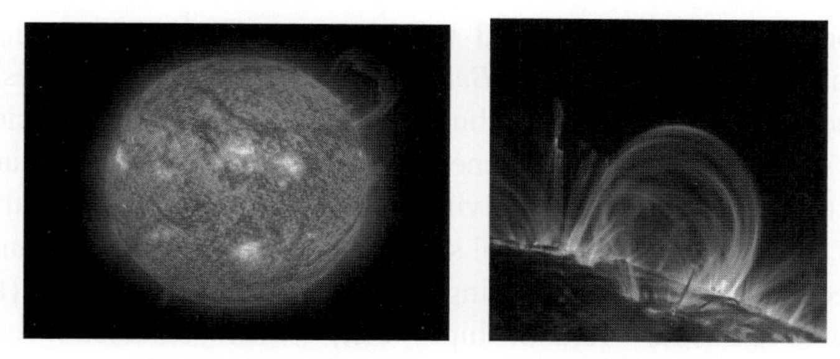

http://www.wingmakers.co.nz/images/sun.jpe/
http://www.wingmakers.co.nz/images/coronaloop_trace_big-browse.jpe

Fig. 11.

Figure 11 shows Sun's surface. Complex and beautiful structures are emerging from random motions of plasma gas which comprises independently moving nuclear particles and electrons. This plasma exists at temperatures of 6000K or higher. The important requirement for such emergence is that the system (Solar surface) be open and that energy be continuously fed into it from the surroundings (bulk of the Sun). Sun has energy aplenty. This simple example demonstrates that in some cases at least, no complex causes (intelligent designers) are required to build complex structures. For example, emergence of fine structure can be observed in a saucepan containing thin layer of oil or water which is slowly heated (see Figure 12). The emerging structures are due to the well known *Benard convection* process.

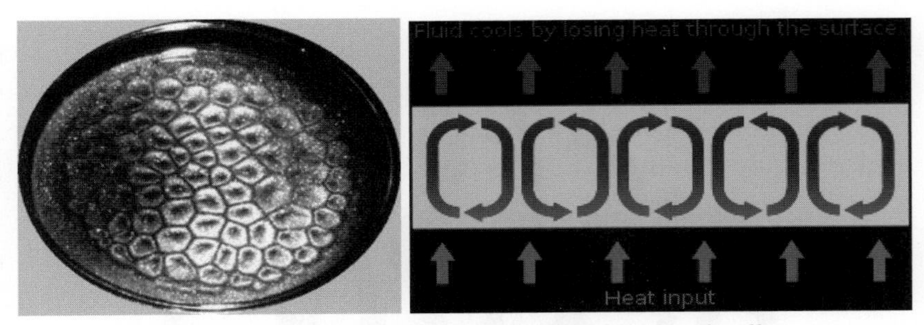

source: http://en.wikipedia.org/wiki/B%C3%A9nard_cell

Fig. 12.

The more energy is transferred to the system the more intricate and numerous emerging structures (*Benard cells*) develop. We digress at this point and propose a simplistic, but intriguing analogy with Society. The analogy suggests that the more energy is transferred to a particular human society (the more advanced civilization uses more energy) the more intricate and fragmented its social structures shall become. We can witness this effect today through breaking of traditional social fabric (families, patriotism, personal relationships etc.) with enhanced economic development. The aim of this comment is not to offer any moral judgment, but to hint at the possible fundamental link between general properties of open dynamical systems and evolving social structures.

Another important requirement for the emergence of complex structures is that parts of the system (e.g. particles) interact with each other via forces which are of intermediate strength. Too strong interactions would preclude generation of fine structures and enforce static regularity. Too weak forces on the other hand would be incapable of holding the complex structures together for any significant length of time.

b. Evolution is just a theory

This objection is often put forward by the opponents of Evolution. Sacred texts in monotheistic religions (especially Islam) have been largely unchanged since they were written many centuries ago. This unchanged form is considered to be evidence of truth when compared to Science which continuously updates and modifies its knowledge.

This objection exemplifies lack of understanding of the scientific method by Creationists. All scientific observations (in contrast to random observations) are theory-laden and there are no "bare facts" in Science. We perform observations with a particular hypothetical concept in mind, not by blindly searching for lucky or any available find.

Theories are necessary because they provide coherent frameworks within which data/facts can be collected, organized and rationalized/explained. The refinement of theories shows how Science continually improves its understanding of the World in response to new information. This improvement is impossible in CS which instead of responding to and adapting to newly acquired information (as Science does), attempts to fit new information into the existing ideas or preconceptions.

Creationist and scientific methodologies

The concept map below juxtaposes and compares the two methodologies. It is interesting that both methodologies start with unproven assumptions. In the case of S it is the 'invented hypothesis', in CS it is the assumption that ET is wrong. The difference arises from the presence of two feedback loops in the scientific and none in the creationist method. This reflects the CS preoccupation with obtaining the predetermined result i.e. disproving the ET. The feedback loops enable Science to be self-correcting and continuously self-improving. CS has no such mechanisms.

The divine (or 'intelligent') designer induced creation of the World is a matter of faith. Faith is not a scientifically verifiable proposition which Science can prove or disprove. The problem arises when it is claimed that observation and evidence lend scientific credibility to the ideas of 'creation science' (CS). One can say with certainty that CS is not a scientific activity irrespective of whether its conclusions are true or not in the final analysis. As already mentioned in Chapter 1, the entities whose existence is postulated outside the material, physical world are not the appropriate subject matter for scientific investigation. CS pretends to be 'scientific' in order to gain respectability which comes with the attribute 'scientific'.

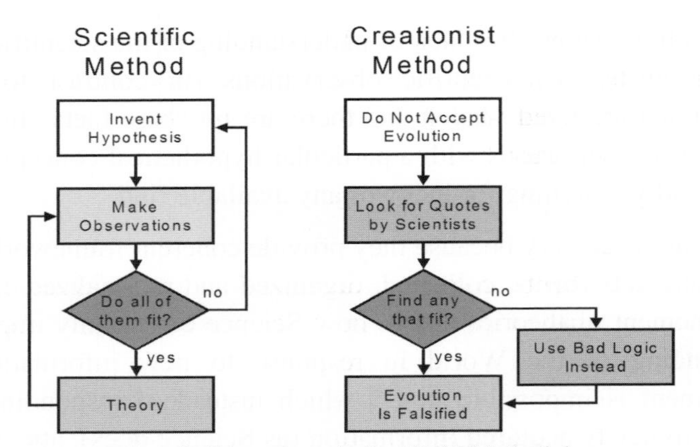

source: http://wiki.cotch.net/upload/9/96/Methodology.png

- CS does not increase our understanding of the World. According to CS the special creation **has occurred** so the evidence gathered is just an afterthought designed to support this preconceived conclusion. CS has no research programme and its 'evidence' consists of poorly correlated arguments against evolutionary theory.
- No mechanisms for understanding the origin of Life are proposed by CS, because the creation idea is 'fixed' and no scepticism is allowed in CS.
- CS has loosely connected, 'ad hoc' thoughts. For example, CS explains the fossil record by claiming that fossils are preserved bodies of creatures which perished in the Great Flood. CS does not explain why then are the creatures not mixed together in the same layer, but instead are stratified in different layers?
- CS disregards the established results on the age of the Earth obtained by nuclear physics (radioactive dating), cosmology (background microwave radiation), astronomy (red shift measurements), geology (erosion, plate tectonics) and does not provide a coherent explanation which will encompass **all** of these results.

For example, if Global Flood indeed occurred, what caused it? Today's floods can be explained by weather conditions. If the whole surface of the Earth was covered in water, where did the water come from and where did it recede after the Flood?

Flood was an act of God; God is not subject to natural laws or experimentation, so claims about God cannot be falsified even in principle. CS is bent on disproving specific explanations given by Science and does not accept that scientific theories must explain **ALL** evidence in a consistent way. Disproving a single scientific result is not sufficient to disprove a whole theory or the scientific method. Furthermore, after disproving one theory an alternative theory needs to be proposed which can consistently explain **all** known results. Scientific knowledge is not a collection of independent facts, but a network of interrelated concepts and empirical results.

The diagram which follows is another typical example of logical fallacy (circular reasoning) employed by CS.

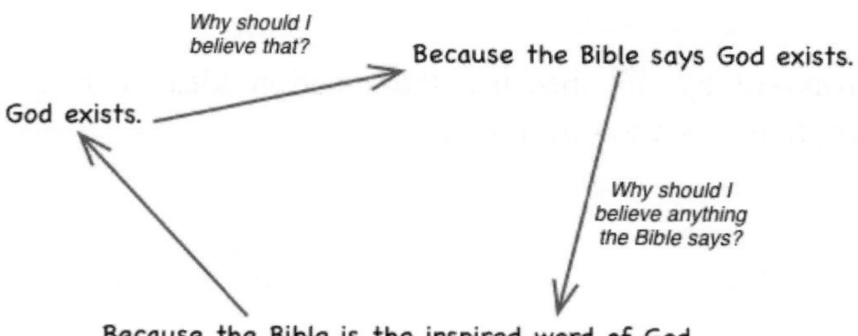

As the quotation from Augustine of Hippo reveals, Creationists understand neither Science nor Religion. They distort the meaning of religious texts and impart exact scientific meaning to texts which were never written for that purpose. Ussher's 'calculation' of the time at which the World was created shows the absurdity of literal interpretation of religious texts for describing/explaining scientific concepts. The absurdity is highlighted further when juxtaposed with Augustine's comment about Time. Augustine describes Time as being tied to the physical objects of Creation. This view is interestingly, in accordance with ideas of modern physics where Time and Matter are seen as existentially dependent entities.

Driving forces of Creationism

Sources of Creationism are manifold. Rapid social change and weakening of social fabric has been stimulated by the growth of Science and Technology. This change is psychologically unsettling and scapegoats for what was perceived as "loss of faith and breakdown of morality" are needed. Creationists blame Theory of Evolution for promoting atheist views. Atheistic extrapolations from Science attacking Religion are also being made, with the result that Creationism continues to exist. The goals of Creation Science are political, not scientific. In USA the constitutional separation between Church and State precludes religious education in state schools. The creationists want their 'theory' to be taught in schools in parallel with evolutionary theory. This may become possible if they can demonstrate that Creationism is a scientific rather than a religious doctrine.

Creationism criticizes the inability of Science to tackle some questions of human meaning and destiny (see quotation below), but it does so in a wrong way i.e. by masquerading as a scientific endeavor which it is not.

A dull future: heat death

Bertrand Russell's acceptance of the heat death (1903)

"...that all the labors of the ages, all the devotion, all the inspiration, all the noonday brightness of human genius, are destined to extinction in the vast death of the solar system, and that the whole temple of Man's achievement must inevitably be buried beneath the debris of a universe in ruins- all these things, if not quite beyond dispute, are yet so nearly certain that no philosophy which rejects them can hope to stand.

"Only within the scaffolding of these truths, only on the firm foundation of unyielding despair, can the soul's habitation henceforth be safely built."

source: http://esmane.physics.lsa.umich.edu/wl/umich/phys/satmorn/2003/20030208/real/slides/img0026.gif

The above quotation from Russell highlights problems which many people have with accepting the scientific view of the World. For many people it is a bleak view (recall the quotation by Flaubert in Chapter 1) and thus unacceptable. However, the bleakness is no excuse for avoidance of truth or for the substitution of truth by something more psychologically palatable. This is an example of 'reality therapy' which Science provides. This 'unacceptability' drives some people into other world views (religion, magic, mysticism). Please note that Russell's view is only one possible interpretation of and the extrapolation from scientific results. Science cannot tell us anything about 'bleakness' or Life's purpose which are subjective, psychological phenomena. Creationism challenges the validity and reliability of scientific knowledge so we shall discuss next what scientific knowledge is, how it is acquired and how its reliability can be evaluated.

Reflection:

Has modern Science disproved Religion? (see Topic 1)

Is there a difference between Creationism and Religion and, if there is, what is it?

Has modern Science contributed to the rise of Creationism and why?

What is the most important principle in Science which can be described as 'Faith'?

Tips: recall from Chapter 1 whether Science can prove/disprove Religion.

Consider the questions which Science cannot answer. Recall the basic principles of scientific method (described earlier) when answering the question about "Faith" in Science.

Scientific Knowledge

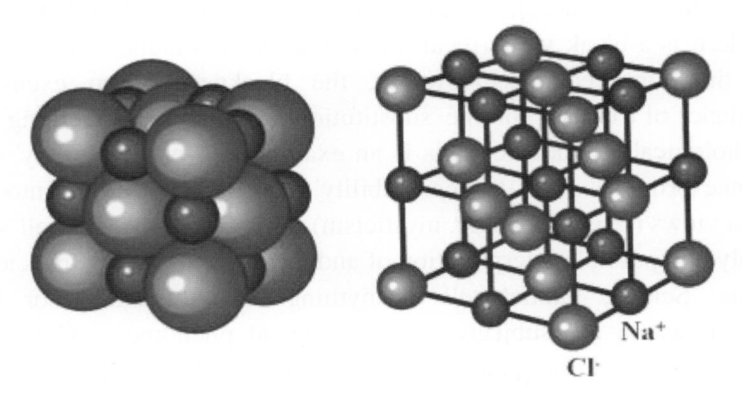

source: http://physicaplus.org.il/11/stones_files/image007.jpg

Fig. 13.

The aim of Science is to understand Nature, so what does it mean to explain, understand or know something?

Let us look at an example. A grain of salt has approximately 10^{16} atoms of sodium and chlorine. Our brain has around 10^{11} neurons. If the position of each atom in the grain of salt is 'recorded' in one neuron, it is obviously impossible to 'know' even a grain of salt. However, there is an **order** in the structure of salt as shown in the drawings of crystal lattices in Fig.13.

The existence of order and regularity in Nature is essential for Science and for scientific understanding. This allows us to reduce, compress and simplify enormous amount of data/observations. We can explain Nature by reference to:

- **causes or causal mechanisms** (e.g. 'Excessive alcohol consumption may cause liver damage.')
- **laws/general principles** (e.g. 'Fuel efficiency of the vehicle is determined by size and weight. This is because according to the

laws of mechanics acceleration is proportional to force, but inversely proportional to mass')

- **underlying processes** (e.g. 'Chest pain is symptomatic of pneumonia. Due to lung infection, alveoli are filled with fluid which impairs the flow of gases')
- **function** (e.g. 'Birds build nests in high places to protect their young from predators.')

The *Covering Law model* of explanation states that explanations are the form of logical argument, they have premises and conclusion [7,14]. This law can be used to describe many, but not all types of explanations. The *Covering Law model* is especially successful if the phenomenon being explained is 'governed' by some general law.

In *Covering Law* formalism premises must be true and entail the conclusion; the premises must contain at least one general law and be relevant to the phenomenon being explained. Explaining the phenomenon implies demonstrating that its occurrence follows deductively from a general law, supplemented if need be by other laws and facts, all of which must be true.

Consider the example which follows and the explanation of the question: *Why is the shadow of the pole 20m long?*

[Light travels in straight lines (General Law)]

⇩

[Sun's elevation=37^0; Pole's height=15m (Particular Facts)]

⇩ Causality

[Pole's shadow=20m (Phenomenon which is to be explained)]

The explanatory relationship is asymmetric. Exchange of the last two statements (below) changes the sequence from an explanation, to a prediction.

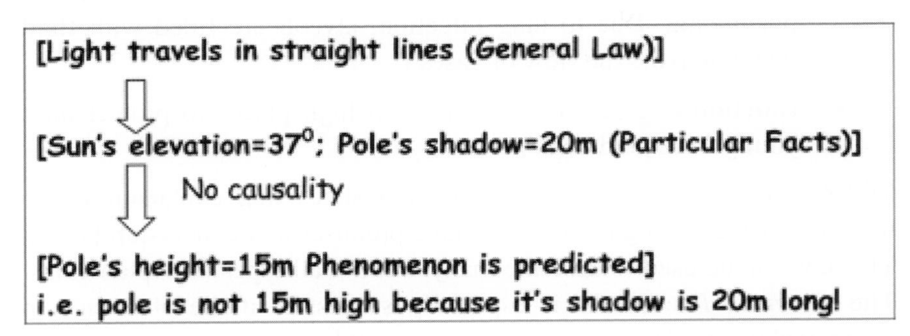

The explanation must contain relevant information in order to be acceptable. In the example below, the sequence given is not an explanation, because the fact that John was taking pills is irrelevant to him getting pregnant.

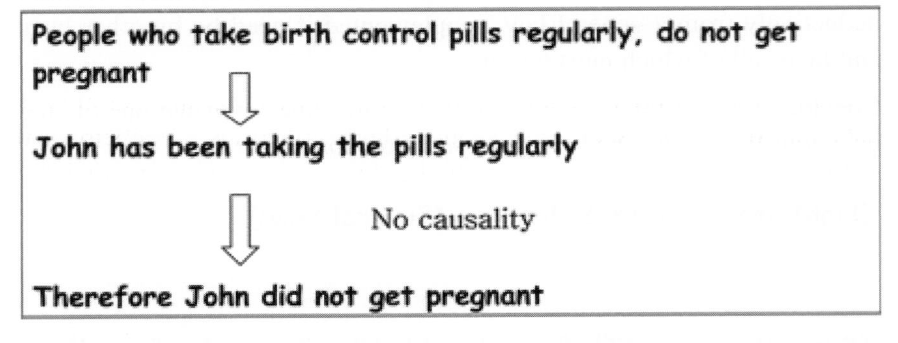

Thus, particular facts and the explanation must be in the causal relationship with each other.

The *Causality-based model* proposes that to explain the phenomenon is to ascertain what caused it. This is a better model because it does not run into fallacies described in the *covering-law model*. However, even this model does not encompass all the explanations. For example, scientists explained that water is H_2O i.e. that one molecule of water consists of two hydrogen atoms and one oxygen atom. However, there is no causality here, a substance being H_2O does not cause it to be water, it just is water!

Causality is difficult to define exactly due to:

- **existence of multiple causes**

What caused my coffee cup to break when I pushed it off the shelf? Me pushing it or the gravity which pulled it downwards? The brittleness of clay which it is made of?

In natural sciences, scientists try to restrict the analysis to situations where a single cause can be identified by giving reproducible, well-known effect.

- **logical necessity**

If the cause occurs, must the effect follow?

Empirically we only see cups being dropped and subsequently them breaking up.

In practice, we often observe only that one event follows another (e.g. night follows day), but that does not imply that they are necessarily causally related (day does not 'cause' night). In Science, the causality is established when scientists devise laws, theories, mathematical models/equations. If one can, with the help of such aids, predict the future events, this shows that Nature is predictable and deterministic. However, we know that many events are random (e.g. throwing of dice). The proof of causal link is often difficult to establish, because many effects are caused by multiple causal factors. While the condition that the cause precedes the effect is a necessary one, it is not sufficient. Correlation is not causation!

A key to cause and effect is the notion that the object's existence and properties are independent of the observation or experiment and rooted in the material reality of Nature. Causal links generate patterns which are manifestations of the Universe's rational order. Does this chain of cause and effect ever end? Is there an `Initial Cause'? God?

The span of scientific knowledge

We shall discuss some fundamental questions regarding scientific knowledge in order to assess how reliable it is. The reliability of scientific knowledge is of course crucially important for Science, but some general aspects of these questions remain unresolved with different thinkers maintaining different points of view.

- Can Science explain everything? (TOE= Theories of Everything) If not, what will remain unexplained?
- Can all other Sciences be reduced to Physics? (Reductionism)
- Is the World deterministic or random? (This has implications for the problems of free will and human destiny)

Some philosophers maintain that it is impossible for Science to explain everything, because explanation always invokes something else. What then explains that 'something else'? Science uses fundamental laws and principles to explain things. Since nothing can explain itself (it would be circular reasoning), it follows that some laws and principles shall remain unexplained. (cf. Gödel theorem)

Example: Will human consciousness remain unexplained or will it be understood in term of exceptionally complex but deterministic biochemistry of the brain?

The outcomes of Natural processes are more complicated than the laws that govern them. Laws possess symmetries which are broken in the outcomes. This is why small number of laws leads to a huge number of complex outcomes, states, structures.

Example: There are only 3 laws of classical mechanics, but there is an infinite variety of motions and shapes that objects assume in reality. A finite number of chemical reactions gives rise to immense diversity of living creatures etc.

This is why even if we succeed in developing theories which encompass all the main forces of Nature (TOE) we would not necessarily be able to predict all possible outcomes or properties. This is because the sub-components of objects can be organized in a huge variety of ways. For example a simple question: "Why are there only nine planets in the Solar system?" has no simple answer even though we have detailed knowledge of the fundamental forces which are active in the Solar system. Some scientists [15,16] have argued that the purpose of modern Science is not to search for 'ultimate laws or explanations' (if such exist), but rather to describe complex, adaptive matter with its emergent behaviour and properties. The emergent properties are due to self-organization of matter and cannot be deduced from 'ultimate laws' or Theories of Everything (TOE). TOEs do not explain every object, physical state, or process, but rather they provide a coherent description of all fundamental forces of Nature.

We know a great deal about atoms of every chemical element, yet the totality of ways in which these atoms can be combined under the influence of energy is beyond our ability to predict with certainty. The combinations of atoms are emerging properties. This is why Science is difficult: we start by observing states or structures of broken symmetry and have to reconstruct the symmetric laws which govern them (we work backwards).

Reflection:

Why should Nature be amenable to human understanding at all?

Is the understanding of Nature an evolutionary necessity for humans?

Tips: consider whether species with such understanding have a survival advantage

How does Science achieve understanding? There are two opposite views about how the World can be understood: *reductionism* and *holism*. Most people hold views which are combinations of the two views (Figure 14 illustrates the difference between *holism* and *reductionism*).

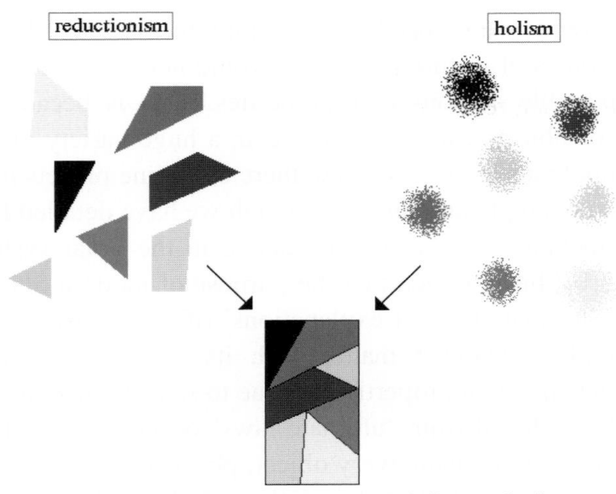

source: http://abyss.uoregon.edu/~js/images/holism

Fig. 14.

Reductionism, determinism and irreversibility

Reductionism is a belief and a methodological principle which claims that any complex phenomena/entities can always be defined or explained in terms of a relatively few simple or fundamental objects or processes.

Reductionism suggests that the breaking down, "reducing" of complex to simpler is the sole path towards understanding. Most fields of Science practice some form of reductionism and mathematical relationships describe the World in reductionist terms.

For example, *atomism* is a form of reductionism because it holds that everything in the Universe can be explained by considering few simple entities (elementary particles) and interactions amongst them which are governed by laws. Modern chemistry reduces most chemical properties of matter to chemical elements and their rules of combination.

To a reductionist, once a set of equations or mathematical relations has been found to describe a system, the behavior of the system is considered to have been fully explained.

The picture of a watch below illustrates reductionism. In order to understand how it works we disassemble it ('reduce it') into composite parts. If we can then reassemble the parts we say that we understand how the watch works. *Reductionism* is a universal principle.

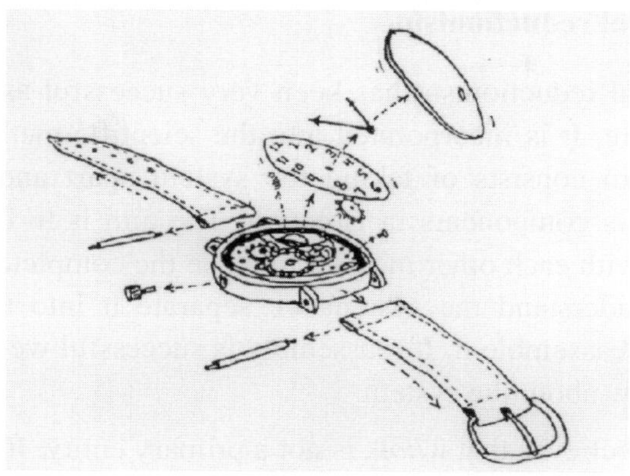

source: http://abyss.uoregon.edu/~js/images/watch.gif

Famous physicist Dirac has opined that since fundamental laws of physics are known, problems in other disciplines e.g. chemistry can be explained solely by (reduced to) the laws of physics [17]. Another famous physicist Rutherford has claimed: "There is physics and there is stamp collecting". Both views are examples of reductionist approach to Science and to the understanding of the World.

Reductionism is related to "Ockham's razor" principle, which states that between the competing ideas, the simplest theory that fits all the facts is the one which should be selected. Reductionism is widely accepted due to its predictive and explanatory power. Reductionism is a good approximation of the macroscopic world but it is not fully adequate for describing the microscopic world, as quantum physics discovered. Reductionism's success has lead to the misconception of *scientism*, the view that Science provides the ultimate truth. Closely related is the view that everything can and should be reduced to the properties of matter (*materialism*). According to reductionism, emotions, beauty and religious experience can be reduced to biological instinct, to chemical reactions in the brain, etc. The 20th century reactions against reductionism are relativism and postmodernism. *Reductionism* can be subdivided in two types: methodological and ontological.

Methodological reductionism

Methodological reductionism has been very successful as a method for studying Nature. It is incorporated into the scientific method. This type of reductionism consists of taking the system apart and studying the properties of its components in isolation. The aim is to determine how parts interact with each other in order to make the complete system work. In order to understand the system we separate it into the constituent elements and reassemble it. If reassembly is successful we have learnt all there is to know about the system.

Reductionist believes that *whole* is not a primary entity. It can be broken down or analyzed into its component parts and the relationships which exist between them. **Holist** maintains that *whole* is the primary entity and is greater than the sum of its parts. Not everything can be fully reduced to the sum of its parts. The Figure 14 illustrates the distinction between holism and reductionism. It is important to note that *reductionism* and *holism* can both be associated with *determinism.* A few examples of reductionism and the state of acceptance of each reductionist description follow.

- Gas laws → reduced to kinetic theory of gases (accepted)
- Genetics → reduced to molecular biology and biochemistry (accepted),
- Psychology → reduced to neuroscience (not yet achieved),
- Psychology → reduced to evolutionary biology (debated).

The Figure 15 of a human face reduced to jigsaw puzzle pieces illustrates the fundamental challenge which reductionism poses for the understanding of our own individuality and social ethics. Can human personality be reduced to mosaic pieces?

source: http://www.thoughttheater.com/

Fig. 15.

Ontological reductionism

Ontological reductionism can be regarded as a belief system and represents the extrapolation from methodological reductionism. The whole is exactly equal to the sum of its parts. There is nothing more to say about the entity besides what can be discovered by this particular (reductionist) scientific method. Ontological reductionism equates the research method with the nature of Reality. There are objections to this type of reductionism (see the quotations which follow).

Physical objects are composed of physical particles (atoms, molecules etc.) and fundamental forces (electromagnetic, gravitational, nuclear force). The particles and forces give rise to complicated structures and emergent properties which cannot be simply reduced to or predicted by the same particles or forces. For example, resin is sticky, but that does not mean that each atom in it possess the property of 'stickiness'.

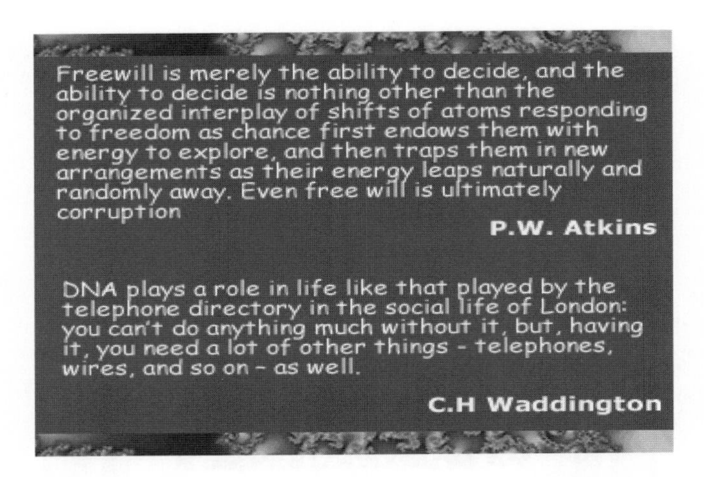

Freewill is merely the ability to decide, and the ability to decide is nothing other than the organized interplay of shifts of atoms responding to freedom as chance first endows them with energy to explore, and then traps them in new arrangements as their energy leaps naturally and randomly away. Even free will is ultimately corruption
P.W. Atkins
DNA plays a role in life like that played by the telephone directory in the social life of London: you can't do anything much without it, but, having it, you need a lot of other things - telephones, wires, and so on - as well.
C.H Waddington

Reduction cannot always involve well defined, one-to-one correspondences between the objects or effects being reduced and entities to which they are being reduced to. We often encounter cases where a single object correlates/corresponds to several other objects or forces and then reductionist approach becomes problematic. The reason for the problem is the existence of interactions between elements of the system which we are trying to describe and reduce. Reduction process destroys such interactions and allows the reduction to simpler entities to be made, but such reduction simultaneously alters the original object (which we are trying to understand) in a fundamental way.

Examples of reductionist statements:

'The sole purpose of a violin is to be the source of sounds of specified wavelength range.'

This scientific description does not rival (contradict) the description of violin as part of the orchestra, as a stimulant for aesthetic feelings.

'The sole reason for existence of organisms is the propagation of DNA.' Biological cells are made up of atoms, but the arrangement of atoms will be very different in different cells at different times. So the concept of 'cell' cannot be defined purely in terms drawn from fundamental physics (mass, acceleration, momentum etc).

Explanation may be possible and necessary at various, complementary levels i.e. the explanations may be complementary rather than exclusive.

Q: When are two descriptions complementary?

A: Only when both are necessary for the full description of reality!

Maps and complementary knowledge

Maps are examples of complementary knowledge. They are representations of reality which contain scientific data. They are not the reality itself, but only a 'slice of reality'. Theories are like contours on the map. They organize individual data or points. Most maps are specialized: geological, population, political, agricultural etc. Two different, but complementary maps of the Earth (poverty distribution map and geographical map) in Figure 16 illustrate the idea of complementary knowledge. Different maps describe the same reality (e.g. Earth), but viewed from different perspectives. They are not exclusive! Likewise, scientific theories or knowledge is only a representation of reality, not the reality itself. It is complementary to other forms of knowledge. It is never full or complete, as *scientism* claims and hopes.

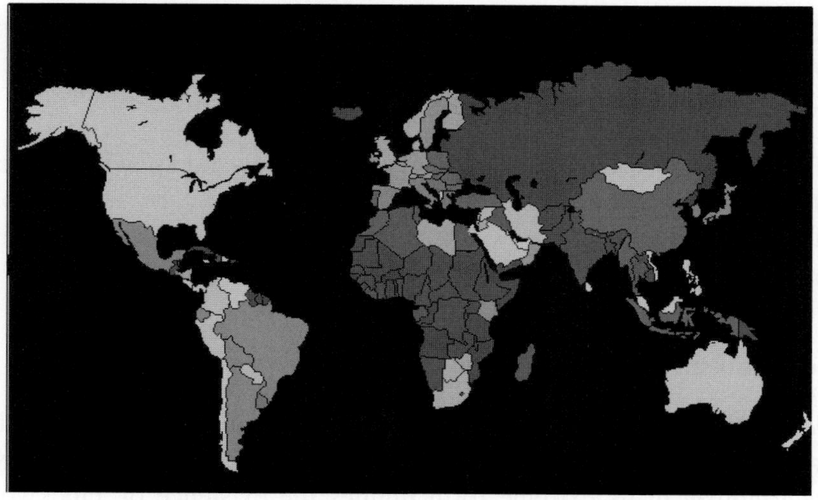

source: http://www.cis.hut.fi/research/som-research/worldmap.html

source: http://www.mapsharing.org/MS-maps/map-pages-worldmap/2-world-map-physical.html

Fig. 16.

Reflection:

Students pay ('exchange') their money (student fees) for knowledge. Explain how this interpretation of knowledge as a commodity, relates to reductionism.

Can the modern idea of 'consumerism' and search for material wealth be related to the general concepts of reductionism?

Why is reductionism so acceptable?

Tips: Consider if the relationship between money and knowledge is of the one-to-one kind. Reflect whether money can be a common denominator? Consider how often the reductionist approach works in everyday life.

Determinism and reductionism

Determinism claims that all natural events are strictly determined by previous causes which are also natural events (see the figure below). *Determinism* in its extreme form implies the existence of 'clockwork

Universe'. According to determinism all natural events can be predicted given sufficiently detailed knowledge of the current state of the Universe and fundamental laws which govern its behavior. It is interesting to note that *reductionism* entails *determinism,* but not vice versa. The modern study of dynamical systems claims that World is *deterministic*, but that it cannot be described via *reductionism* [18-20]. *Determinism* does not *necessarily* imply *atheism* either, yet many determinist thinkers are atheists. However, the role of God is severely limited in a deterministic Universe where he serves only as the initiator of Creation or 'initial condition' (Figure 17). After creation, the Universe is left to evolve according to preset laws and God has no further role to play.

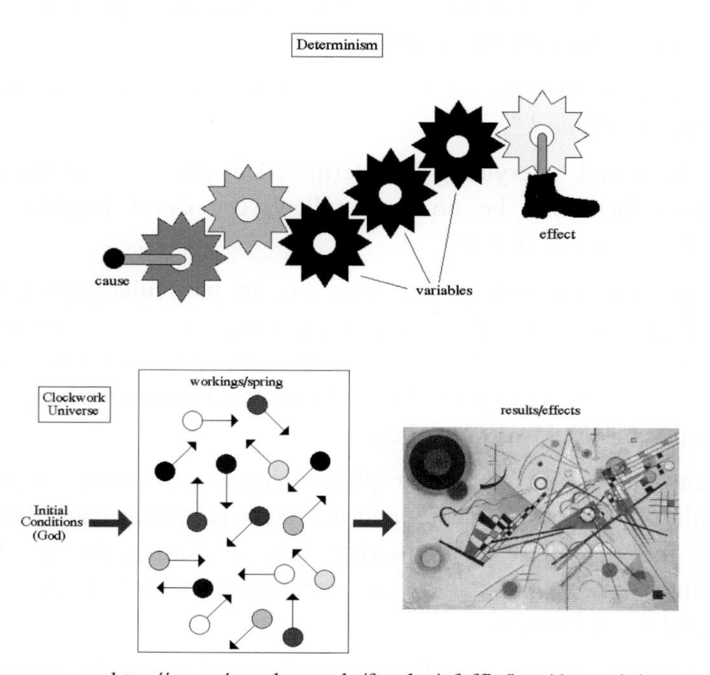

sources: http://www4.carthage.edu/faculty/pfaffle/hgp/determinism
http://abyss.uoregon.edu/~js/images/clockwork_universe.gif

Fig. 17.

Determinism claims that all events, including moral choices, are completely determined by previously existing causes. That precludes free will and excludes the possibility that humans could have acted in a different way from the way they did. *Determinism* holds that Universe is fully rational because complete knowledge of any given situation assures that accurate knowledge of any future development is also possible. The present state of Universe is fully determined by its previous states and the present state is the cause of the states which will follow. If Mind, at any given moment, could know all the forces operating in Nature and the respective positions of all its components, it would thereby accurately know the future and the past of every entity in the Universe. (This argument leaves aside the practical issue of whether such Mind can actually exist!) The Persian poet *Omar Khayyam* expressed a similar *deterministic* sentiment by saying:

"And the first Morning of Creation wrote, what the Last Dawn of Reckoning shall read."

As the poet said, everything is determined in advance and under such conditions there can be no free will. Every event happens out of necessity, not out of choice.

Scientific laws are strictly deterministic. In particular, Newtonian or classical physics is rigidly deterministic (both in the predictions of its equations and in its foundations) and leaves no room for chance, surprise and creativity. According to classical physics, everything happens as it must happen ('clockwork Universe').

To recap, *determinism* as the philosophical framework claims that everything has a unique cause, and that each particular cause leads to a unique and distinct effect. Another way of stating this is that for everything that happens there are conditions such that, given them, nothing else can happen.

Indeterminism, on the other hand, though not denying the influence of behavioral patterns and extrinsic forces on human actions, postulates the existence of free will and denies rigid predetermination of events. Supporters of determinism defend their theory as compatible with moral responsibility by arguing, for example, that evil results of certain actions can be foreseen and this in itself imposes moral responsibility and creates a deterrent for certain human actions.

Reductionism and *determinism* when coupled together inevitably lead many thinkers to reject any non-material causes or effects in the World and espouse *materialism*. These thinkers also reject religion and adhere to *atheism*. To atheists all explanations of entities and processes in the World must be sought strictly within the World itself. The arguments about reductionist view of the World and the extrapolations which are deduced from it are (and have been) subjects of extensive debates. The supporters of atheism have even resorted to advertising campaigns (Figure 18). The issue at stake is important because it influences decisions about how best to live our lives and organize our societies. The groups of atheists have recently commissioned a series of adverts on public buses in several countries which proclaim their views and by which they hope to counter dangerous (as they see it) spreading of religious beliefs. This may sound amusing, but it also demonstrates how questions extrapolated from scientific arguments deeply affect human lives. The atheist advert correctly uses statistical argument against God's existence ('There's probably no God…') rather than a definitive claim, because God's existence cannot be proved or disproved (see Chapter 1).

The main problem for *reductionism* is that many (though not all) processes at macroscopic level are irreversible. On the other hand processes at microscopic level are reversible. If processes are irreversible then they cannot be easily 'reduced' to simple ones, because operation of reduction would entail the loss of processes' vital components. Reduction of the final result to the initial state/cause cannot be achieved if we cannot perform a re-assembly! Examples of reversible processes occur in mechanical devices e.g. rotation of motors. Examples of irreversibility abound and include the prominent ones like human Life. We age irreversibly (in spite of various cosmetic treatments) and eventually die. We are 'imprisoned' in the 'time capsule'. The sand clock, like our life runs irreversibly forward towards its final state

Fig. 18.

(Figure 19). However, the notion of irreversibility cannot be used to support religious beliefs by denying *materialism* and *determinism*. This is because the process may be deterministic and irreversible, as modern physics (statistical mechanics) demonstrates [20].

source: http://www.columbia.edu/cu/seminars/death/DeathSeminarIndex.html

Fig. 19.

Holism

Holism is the view opposite to reductionism and suggests that there are 'emergent properties' in a complex dynamical system that cannot be predicted solely by analysis of the components of the system. Holism is criticized by reductionists as disguised *vitalism* (proposition of the existence of mysterious life-force) and idealism which can lead to religious thinking.

Study of complex systems sheds some light on these fundamental, philosophical, unresolved questions. Research has discovered that although the equations governing physical events are purely deterministic and time reversible (i.e. the same equations govern the system's evolution forwards and backwards in time), many complex systems are so sensitive to initial conditions that they may evolve along diverging paths and thus exhibit unpredictable behavior. Unpredictable behavior leads to irreversibility.

Figure 20 show how the system may evolve from the initial to the final state along a deterministic path. If the system responds *linearly* to external influences, the final and initial states will be similar (light and dark regions have similar sizes) and no qualitatively new phenomena will be observed. Such process is reversible. If the response is *nonlinear* the system's final state shall be very sensitive to initial conditions and very different from the initial state. On the diagram below, the initial state region has 'disintegrated' into a widely spaced array of points (final states). This entails irreversibility. There are many examples of linearity and nonlinearity. For example, if I double the force available to lift a particular weight I can lift twice the initial weight (linear system). However, if I double the amount of money I earn I do not double the extent of personal happiness (nonlinear system).

a linear or deterministic system is one where initial conditions lead to unique final states, and trajectories that start at similar points, end in similar regions

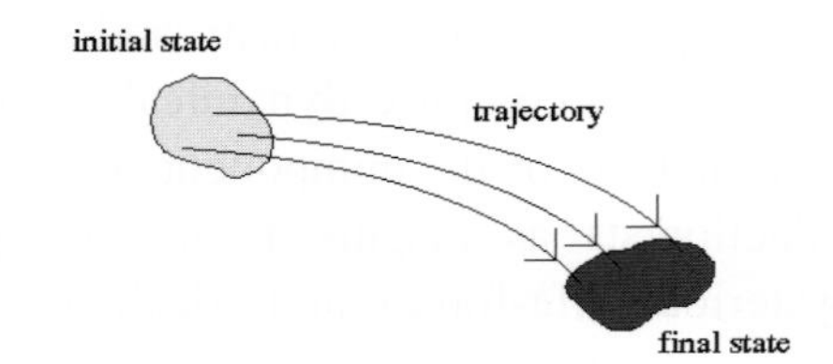

non−linear or chaotic systems are ones where the initial states lead to unique final states, yet the trajectories are not similar even when the initial states are closely spaced

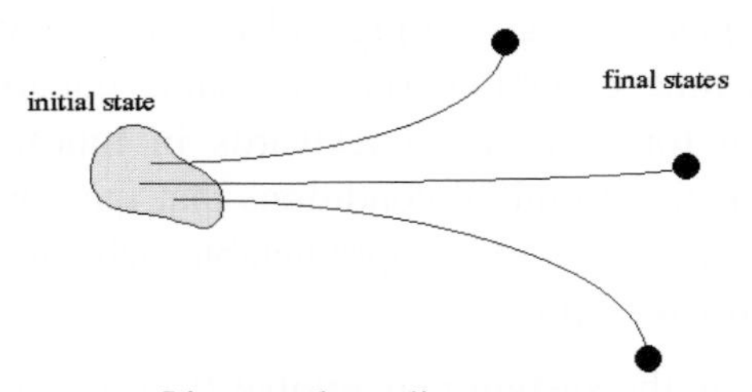

Linear and non-linear systems
source: http://abyss.uoregon.edu/~js/images/nonlinear.gif

Fig. 20.

Reflection:

Determinism has often been criticized for seemingly denying the existence of human free will. If everything is predetermined, then I cannot have a free will. Does the opposite view of Nature as being subject to random events (indeterminism) support the existence of free will?

Tip: consider what indeterminism implies for human choice and therefore for the free will! See ref.[12], pp.134-137 for more detailed discussion.

A general comment regarding reductionism and holism can be made. Both views are based on extrapolations, but out of necessity rather than out of choice.

Humans have natural propensity to extrapolate their knowledge i.e. to use what they know about as an explanation tool for what they do not know about. The need to know has its roots in the necessities of physical survival. (If I am unfamiliar with an object it may pose a threat.). Extrapolation is an application of proven knowledge outside the range of its known validity and is closely related to the 'problem of induction'. This implies that I use what I know to be true under a given set of conditions, under a different set of conditions. If the two sets of conditions are not 'too different' I may be justified in doing so. What is 'not too different' though? Since I cannot know all imaginable conditions I am forced to extrapolate! This is why in Science we need to combine deduction and induction in order to make progress. Fortunately, experiment/observation is our check on whether the extrapolation is justified or not. However if we cannot perform the experiment/observation on a particular phenomenon/concept we lose the feedback loops shown earlier in the concept map of scientific method and thus leave the realm of Science. In short, our extrapolated knowledge becomes uncertain.

Scientific knowledge and Philosophy of Science

Philosophical analysis probes implicit assumptions made by scientists which they themselves are not explicitly aware of. Scientists are fully preoccupied by daily practice of Science and use scientific methods which they acquired as part of their professional training, without reflecting on the methods learnt.

- How do we acquire knowledge about the World, what does it mean to know, what are the sources of scientific knowledge & role of Science?
- What is the nature of Reality?

These questions are answered in different ways by different groups of philosophers. We shall mention some attempts to answer these questions [14].

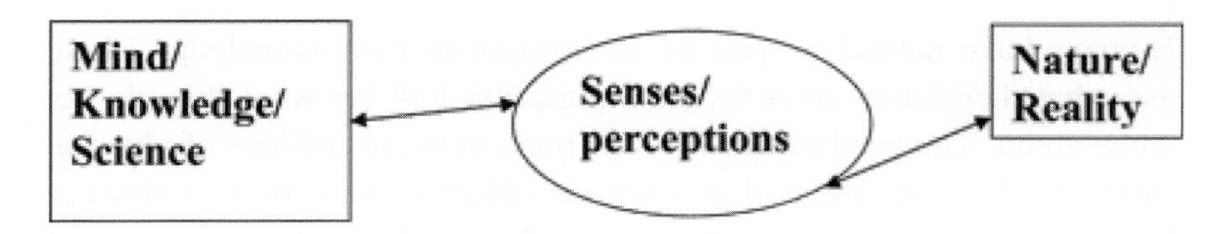

Empiricism claims that all knowledge comes from experience and that this empirical knowledge should be the basis of Science. Scientific activity should involve systematic gathering of empirical facts (measurements, observations).

History of Science has shown that knowledge cannot be solely based on empirical facts i.e. we need to develop theories & models. 'Problem of induction' represents an obstacle for empiricism. Empirical observations alone are insufficient to prove the truth of a conclusion.

Rationalism is a reaction to empiricism. Senses can be fooled (optical illusions demonstrate that) so knowledge and Science should be based on Mind through the exercise of deductive logic and mathematics. Extreme rationalism suggests that Truth can be comprehended directly by the Mind (bypassing the need for empirical data), by proper thought processes starting from the 'first principles'. This is patently untrue because we do need experiment/observation as a check on whether our mental concepts correspond to the World around us or not.

Neither *empiricism* nor *rationalism can be* adequate philosophical and methodological foundations for Science.

Logical positivism states that only directly observable objects and events are valid subjects of scientific investigation. The task of Science is to ascertain logical relationships between observables. Logical connections between observables, based on mathematical logic, represent a scientific theory. This theory is the only "positive" knowledge that we can have. Problems with logical positivism include:

- many objects encountered in Science are not directly observable,

- facts are theory-laden and observation is not a simple act which directly relates observation to Reality
- problem of induction (the story about the turkey being fed at 9am every day!) restricts our knowledge. This is because inductive method operates on observables only and the set of available observables is always incomplete. Theories and hypotheses would have to be excluded from Science if we follow *logical positivism.*

Observation is not a simple act; it is a result of a complex sequence of activities.

Example: seeing an object implies: light scattered from the object → has different wavelengths/intensities → passes through optical elements of an eye and reaches retina → the neurons are fired → their impulses reach the brain → brain interprets & assembles all the inputs.

Figure 21 describes optical illusion and shows that senses alone can be deceived and that the information received by senses is processed by our brain.

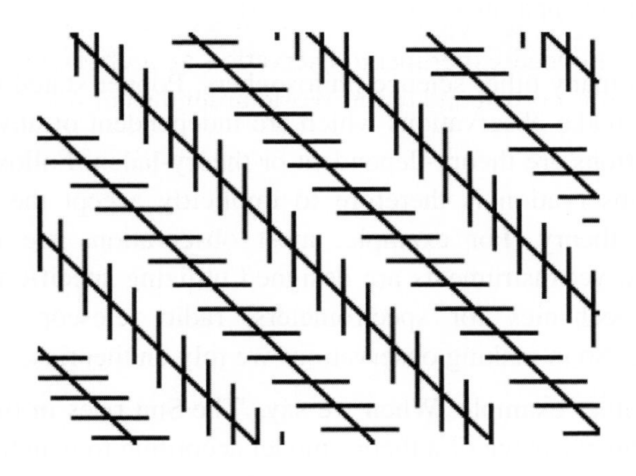

Are these lines parallel?

Fig. 21.

Reflection:

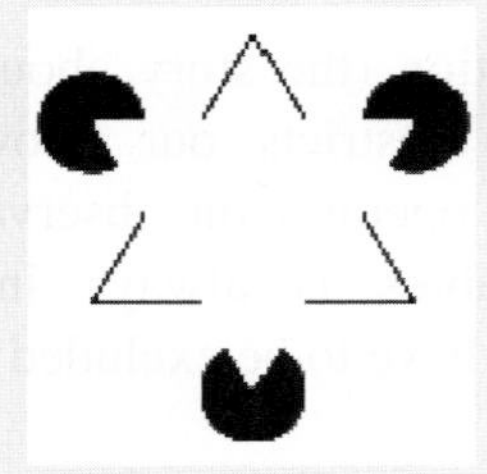

Consider optical illusions (e.g. the one above) and then consider the quotation from W. Heisenberg:

"Natural science does not simply describe and explain nature; it is part of the interplay between nature and ourselves; it describes nature as exposed to our method of questioning."

Optical illusions demonstrate the importance of the role of observer. How and under what circumstances does the observer influence the result he observes?

Tip: Consider relative sizes of the object of observation and the size of the measuring apparatus. see also ref.[14]

Along with many other science philosophers, Popper stated that it is not possible to make observations which are independent of any theory, i.e. all observations are theory-dependent or theory-laden. Following this, to make an observation is therefore to implicitly accept the truth of an underlying theory. For example, most observations are made using instruments, yet instruments are designed utilizing specific theories e.g. quantum mechanics for spectrometers, radio telescope or electron microscope. So in making observations we rely on theories.

Here is another example. When we say, 'The Sun rises in the morning!' we imply the existence of a theory/model according to which the Earth is stationary and the Sun moves around it (the geocentric model of the solar system). Actually, we should say, 'The Earth is spinning!' in order to conform to the heliocentric model.

If facts are theory-laden (interweaving theory and observation) then scientific theories cannot be proved by purely empirical means, because that would imply circular reasoning.

The question is to what extent theory-laden facts affect our ability to validate scientific theory and ensure the reliability of scientific knowledge. Does scientific knowledge faithfully represent Reality? The cartoons below illustrate some of these points.

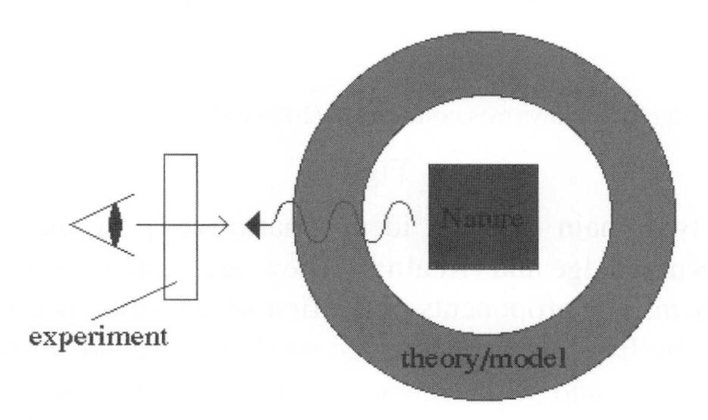

We often 'see' Nature and process empirical information through our own theoretical frameworks as the example of rising Sun demonstrated (see Figure 22).

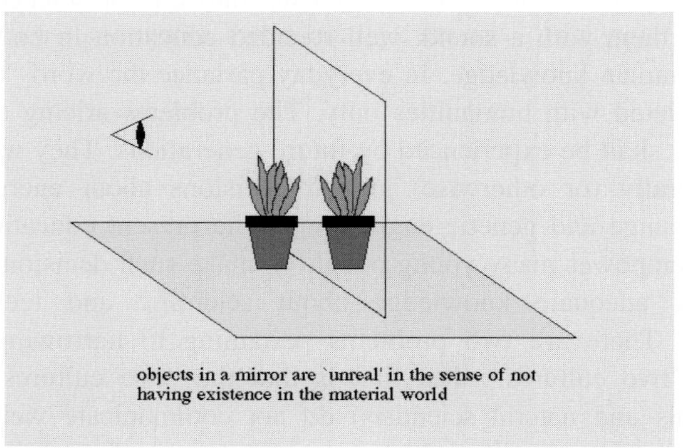

source: http://abyss.uoregon.edu/~js/images/mirror_reality.gif

Fig. 22.

source: http://graphics8.nytimes.com/images/2008/05/26/science/27angi02_650.jpg

Fig. 23.

There are two main sets of ideas ('narratives') in modern Culture regarding Knowledge and Reality. They are called *modernism* and *postmodernism*. The proponents of the first set are often natural scientists (represented by Einstein in Figure 23) while the second set is often held by thinkers in arts and social sciences (represented by Shakespeare) who comprise the bulk of leading social elites. This dichotomy has been first highlighted by C.P. Snow in 1959 and labeled "two cultures". The dichotomy persist today [21] and is reinforced by overspecialized education system which streams students into arts or sciences without providing them with a sound, well rounded education in both of these fields of human knowledge. In everyday parlance the word "culture" is still associated with humanities only. The problems arising out of this dichotomy shall be experienced by future generations. They will have to make morally (or otherwise) guided decisions about energy policy, climate change and genetic engineering. The present education system does not empower many young people to make such decisions, because they lack adequate knowledge about scientific and technological questions. There are two problems pertaining to narrowing the gap between "two cultures". The first is that the "two cultures" (literary intellectuals and natural scientists) do not communicate well and this weakens human resources which are needed to confront arising global problems. Some progress has been made in bridging the two cultures gap by modern information technology, but this progress was 'imposed' by technological development rather that originating from the 'meeting of minds' between two groups of thinkers. To make matters worse even the communication within sciences themselves has become more difficult as Science splits into sub-cultures/sub-fields. The second problem is the asymmetry in the distribution of social influence between the two cultures. Political and business elites which lead Societies have little understanding of ST even though we live in the Age of Science & Technology. This book is written in an attempt to bridge and reduce this dichotomy.

1. Modernism

Modernism originated in 19th century Europe and is today synonymous with the Western World and its predominant global influence via consumer oriented, market capitalism. Modernism asserts that there are ultimate principles (scientific, philosophical) which can explain everything in the World through the intervention of human reason. Reason is the ultimate judge of what is true, right or good. Freedom means successful adaptation to the laws of Nature and of Society which in turn are discovered through Reason and experiment. In other words modernism follows the equation: rational=truth=good=right.

Modernism postulates that Reality exists independently of the observer. The Reality can be understood/known objectively only through Science which provides universal truths. The knowledge produced by Science will always lead to social progress/perfection. Science is a paradigm for all socially useful knowledge; it is neutral, objective; scientists produce knowledge through their unbiased rational capabilities. Modernism may lead to the misconception and oversimplification of 'scientism' or to ethical blunders of 'naturalistic fallacy' (see Chapters 1 and 3, respectively). Modernism is related to realism and materialism as philosophical descriptions of Reality and world view.

Realism asserts that physical world exists independently of the observer and his perceptions. Guided by human reason which analyzes perceptions and empirical information, we can get closer to true Reality and be able to describe it reasonably well. Most scientists are 'realists'.

2. Postmodernism

Postmodernism is a reaction against the assumed certainty of scientific, generally valid, objective picture of Reality. *Postmodernism* recognizes that Reality is not simply mirrored in the human mind, but is actively constructed/interpreted by the brain/mind as it tries to gain understanding. Sometimes this view is called constructionism. In this sense, *postmodernism* is related to *relativism* as the world view.

> *"Natural science does not simply describe and explain Nature; it is part of the interplay between Nature and ourselves; it describes Nature as exposed to our method of questioning."*
>
> *W.Heisenberg*

Relativism overemphasizes the role of social factors in the development of Science ('scientific revolutions'). It also claims that because data are 'theory-laden', Science follows circular reasoning and cannot arrive at the Truth. Science is thus a problem-driven mechanism which can solve problems, but does not explain objective reality. One of the most prominent physicists Heisenberg highlighted the role of the observer in our attempts to obtain scientific knowledge (see his quotation above).

Reflection:

Do you believe in the existence of atoms and electrons?

On the basis of this answer are you a scientific realist or anti-realist (constructivist) in your world view?

Do atoms and electrons exist as we describe them or are they just a convenient mathematical construct that allows us to predict phenomena and properties of matter? See Figure 24.

Tip: to answer the question about the reality of atoms, consider the observed images below and decide how you could prove whether they are false or not? If you cannot think of a proof what would that suggest? see also refs. [14,21]

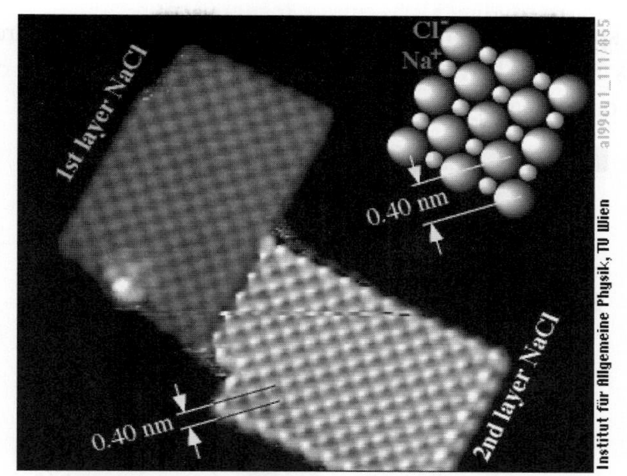

source: http://upload.wikimedia.org/wikipedia/commons/thumb/f/f9/
ScanningTunnelingMicroscope_schematic.png/400px-
ScanningTunnelingMicroscope_schematic.png

Fig. 24.

The Heisenberg quotation may appear at first glance to give support to postmodernist or antirealist ideas. It does not. When different observers perform measurements of properties of e.g. the atom of helium they will get the same, reproducible results. The use of different experimental methods also gives the same atomic properties.

What is the source of these dilemmas about the reliability of scientific knowledge? Objects and events in Nature span vast spatial and temporal ranges/scales (see figures below). The differences are of the order of billions of billions of times. Evolution has prepared our senses and brain primarily to deal with events and objects which are crucial to our immediate survival. This is why not everything can be observed and subsequently understood directly by human senses or minds; this is especially so when it comes to very large and very small objects or to very short or very slow processes. This necessitates the use of aids (instruments, theories) in the acquisition of scientific knowledge. There is however consistency and reproducibility in this acquired knowledge which mitigates against the notion that scientific knowledge is only human construction.

Ranges of spatial and temporal scales of objects and events are shown in the Figures 25-26. The ranges are enormous with human size or life span occupying only a barely detectable interval.

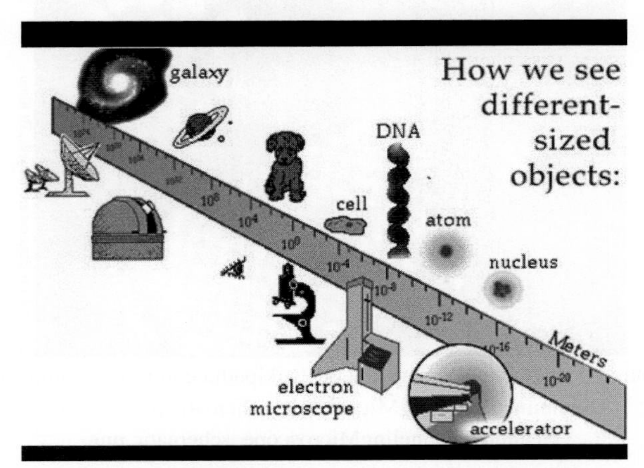

source: http://abyss.uoregon.edu/~js/21st_century_science/lectures/

Fig. 25.

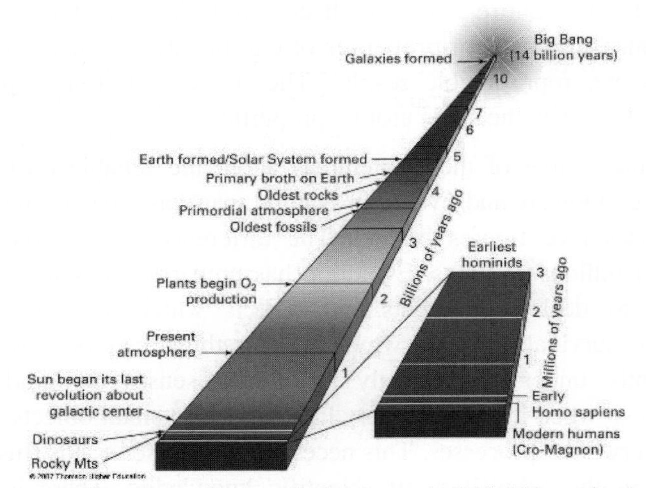

source: http://www.ifa.hawaii.edu/~barnes/ast110_06/sosat.html

Fig. 26.

How can we validate/select a scientific theory?

source: http://huizen.daxis.nl/~henkt/plaatjes/popper-karl-01.jpg

Fig. 27.

The problem of induction complicates scientific method and the validation of theories and had been addressed by many philosophers of Science [14]. The validation of theories is still an active area of debate.

Karl Popper (1902-1994) (Figure 27) was associated with logical positivists and proposed the falsificationist method as the way out of induction problem. **Falsification** is the attempt to show that Science is rational, yet it doesn't need to use induction method to validate knowledge. This view has led to much criticism of falsificationism. For example, is good corroboration a reason to support a theory? How do we know when a test is rigorous and conclusive?

Popper has proposed a demarcation criterion for distinguishing genuine Science from Pseudoscience. He proposed falsifiability as the criterion of demarcation. That is, genuinely scientific theories must be falsifiable at least in principle. For every theory there must exist a test method and an outcome which would refute it. For this reason, Popper held that Marxism and Psychoanalysis are unscientific because they appear unfalsifiable. The collapse of communism in 1989 may have shown that Marxism is falsifiable after all!

Popper suggested that we should accept a Theory on a provisional basis (accept it as long as it continues to be confirmed by experiments). We cannot claim that Theory is absolutely correct, only that it has been verified. However, when an instance is found when the prediction of the Theory is not true, the Theory is falsified and we can claim conclusively that the Theory is wrong. The theory which makes more statements that can be refuted i.e. empirically tested is better than the one which makes fewer such claims!

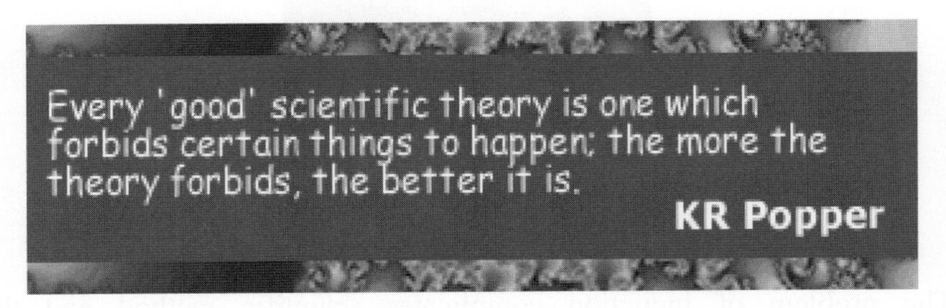

Every 'good' scientific theory is one which forbids certain things to happen; the more the theory forbids, the better it is.
KR Popper

The diagrams of two cycles shown below are good descriptions, according to Popper, of the ways in which Science operates. One may start with the aim of proving (verifying) a theory so the verification cycle applies. In this case one starts at 'Facts' and ends at 'Facts' point in the cycle. However, the two sets of 'Facts' are not identical!

For example, I start by observing that all known metals expand on heating ('facts'). I then formulate a theory that suggests all metals expand on heating. I discover a new metal. My theory predicts that this new metal should also expand upon heating. I perform an observation (measurement) by heating the new metal which confirms my theory. This measurement on a new metal thus also becomes a 'fact' in the verification cycle, but not the same as the initial one. The verification cycle is commonly used in scientific activity, but very large number of individual cases is required to verify a Theory with any certainty. One should pass all (infinite number?) of the cases through the cycle but this is impossible.

Verification cycle

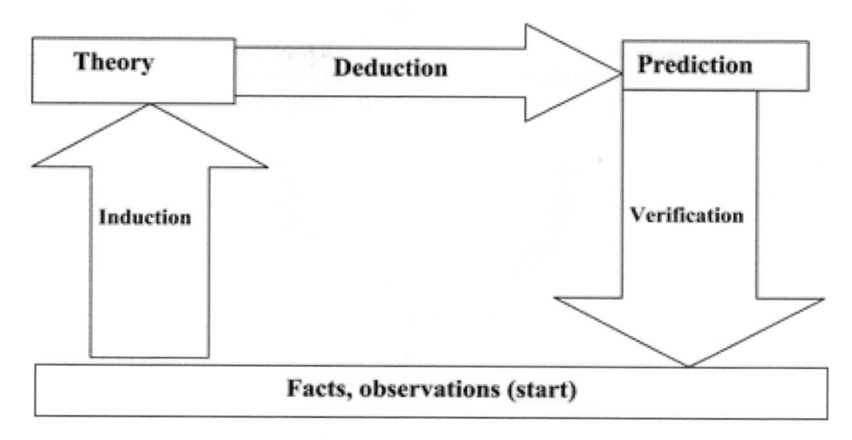

Falsification cycle

Popper has suggested that, because of the 'induction problem' which requires infinitely large number of 'facts' to prove a theory right, we should reason differently. Scientific theories are general and finding a single case (or a finite, small number of cases) which does not fit the theory would disprove it. In this way we can much more easily get rid of theories which are not correct.

The brief explanation of falsification cycle is given below, accompanied by two diagrams. Let us assume that a scientist develops a new theory which says that swans cannot have blue feathers. I disagree and want to prove him wrong. All I need to do is to find a single blue swan and his theory would be falsified and rejected. In the falsification cycles shown below one starts from the 'theoretical proposition' point in the cycle (conjecture that there are no blue swans) and checks the colour of swans (empirical data examination). This is an 'observation' element in the cycle. If at this stage of the cycle one discovers a blue swan, this data can be used to refute the Theory and complete the cycle. One is then ready to start falsifying another theory and begin a new cycle.

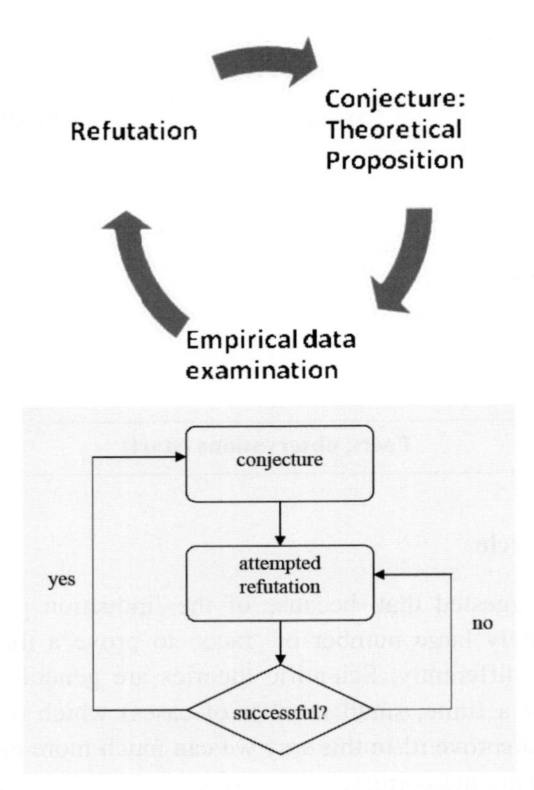

If one attempt at refutation fails one tries refutation again using different arguments or data.

Falsificationism is an example of reductionist approach. There are however problems with falsificationism which prevent it from being used as an absolute, universal principle for determining the validity of scientific theories.

Auxiliary hypothesis. What kind of observation counts as falsification? A theory which is disconfirmed can be salvaged by adding an auxiliary hypothesis – i.e. the original assertion was not wrong, the error was in one of the background assumptions.

Example: would a genetically engineered blue swan prove or disprove the above "swan theory"?

Example: The discovery of planet Neptune. Planet Uranus was found to have anomalous orbit, not conforming to the Newton law of gravitation. However, Newton law was found to be valid for a large number of diverse observations and scientists did not want to abandon it. They introduced auxiliary hypothesis instead which stated that the behavior of Uranus is due to the presence of as yet undiscovered planet which interacts via gravitational force with Uranus and thus causes the anomalies. The planet was subsequently discovered on the basis of prediction made by Newton gravitational law and was called Neptune.

Quine-Duhem thesis [7,14] is related to this problem and states: "The theory cannot be conclusively falsified because it is possible that some part of the complex test situation, other than the theory itself, is responsible for erroneous prediction." To disprove a theory one must rely on other theories, which in turn must be true. If other theories are at fault the falsification becomes invalid. Because experimental and theoretical entities in Science are highly interdependent, it is difficult to formulate an unambiguously falsifiable statement (this reflects holistic nature of human knowledge). However, the adjustments needed to save the theory, while logically possible may be unacceptable on other grounds. Quine-Duhem thesis does not imply that all theories are equally good, but only that theories are underdetermined by facts i.e. we often do not have enough facts to conclusively prove or disprove a Theory. Quine-Duhem thesis does not support relativism.

For example, to claim that the anomalous orbit of Uranus is due to random causes is not acceptable, because the question would then arise why are effects of randomness small and why they do not appear in the orbits of other planets.

Popper has stipulated that, once promulgated, the theory should not be changed in response to experiments which disagree with it (auxiliary hypothesis). One cannot hope to hit the constantly moving target! Only those changes are acceptable which do not diminish, but rather increase falsifiability/testability of the theory. This suggestion is logically clear and an example of 'fair play', but in practice scientists do not always follow this prescription. 'Testability' refers to the testability available with the current state of knowledge.

Do Scientists attempt to falsify theories?

Falsificationism describes how scientists ought to act, not how they do act in practice. Scientists usually look for ways of proving their theory correct, not the opposite! This attitude may not be as detrimental to development of Science as it seems [22]. The example below describes investigation of the structure benzene molecule and illustrates issues regarding scientific practice and falsificationism.

The structure of benzene molecule. Benzene was known (from experiments) to have the formula C_6H_6. There are several conceivable molecular structures which are consistent with this formula and which respect the fourfold valency of carbon atom.

The theory: Kekule proposed in 1865 the molecular structure of benzene molecule in the shape of a hexagonal ring.

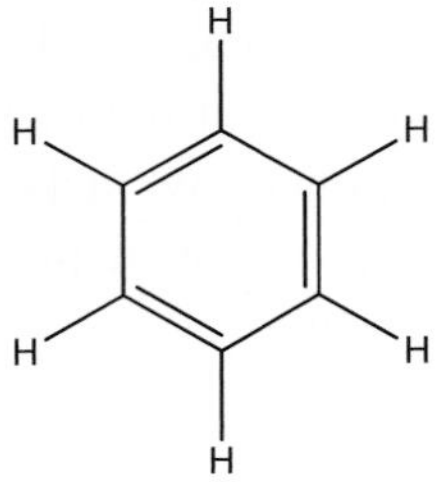

Verification: The number of distinguishable isomers of poly-substituted benzene derivatives (when a hydrogen atom is replaced by other atoms or groups; X) corresponded to Kekule's prediction. In Kekule's time there was no X-ray diffraction or spectroscopic method available to directly probe the molecules. All that was possible was to prepare different derivatives (compounds) of benzene and test their identity by measuring density, melting point or boiling point of each derivative. Counting the number of isomers gave a clue about whether all hydrogen atoms are equivalent or not.

Alternative theory: Ladenburg proposed in 1869 an alternative structure for benzene molecule called prismane.

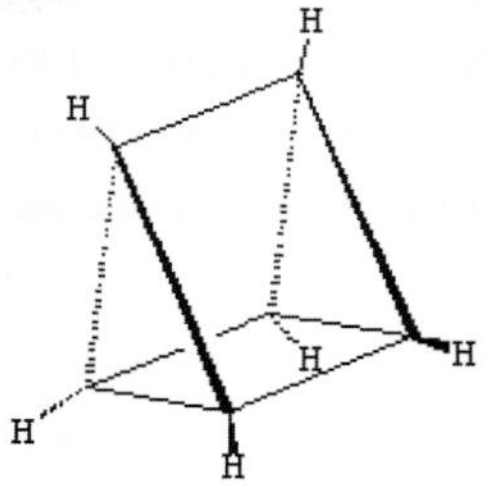

Ladenburg pointed out that there should be two different di-substituted structural isomers of benzene (shown below), but only one isomer was found experimentally.

Auxiliary hypothesis: Kekule subsequently proposed 'ad hoc', auxiliary hypothesis to salvage his original theory. His auxiliary theory could have had several forms. He did not claim that the properties of isomers were so similar as to be experimentally indistinguishable. Since the non-existence of the predicted number of isomers was used as verifying tool (negative criterion), Kekule could have claimed that other isomers will be found in the future. He did not!

He proposed that all C-C bonds are equivalent and that the equivalence occurs via collisions of carbon atoms within the molecule. These collisions happen non-randomly, in a pre-assigned sequence! This was a very daring proposition, since it was not based on any contemporary theory or experiment; it was neither understood nor falsifiable/testable at the time.

It was a 'hypothesis of embarrassment', but its application lead to enormous progress in organic synthesis and to the discovery of many new chemical compounds. Chemists thought that for whatever reason, Kekule's theory must be right and carried on applying it, furthering their research. Kekule model was only fully understood 60 years later within the quantum mechanical theory of chemical bonding and resonance. The true description of benzene consists of two complementary resonance structures shown below, both of which are necessary to convey the true picture of benzene.

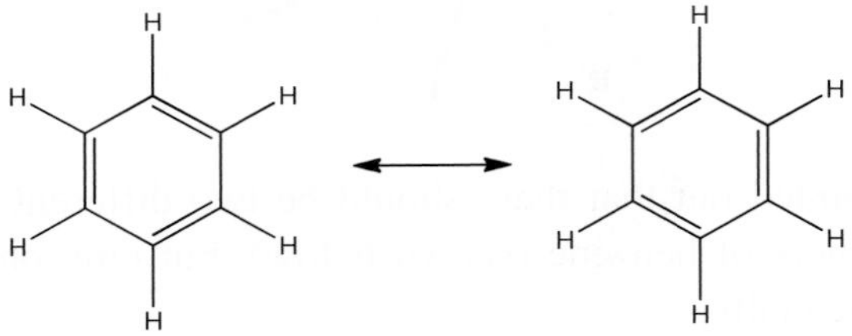

The discussion of problems related to Popper's ideas and their relevance for practicing scientists continues to this day. An example of these discussions is the article published by Buskirk and Baradaran [23] regarding whether the mechanisms of chemical reactions can be proven or not. The important point which this article brings out is the nature of what constitutes a proof. Are we talking about the absolute (mathematical?) proof or "proof beyond reasonable doubt"? Reaction mechanisms cannot be proven in an absolute sense due to the theory-laden nature of data and the necessity of inductive reasoning in Science as discussed earlier in this Chapter.

Models of development of Science

There are three useful models which describe development of Science:

- scientific revolutions
- jigsaw puzzle
- knowledge filter

Each of these models captures some, but not all characteristics of Science. The three models complement each other.

1. Scientific revolutions

source: http://placeforfuture.org/upload/Snimki/Kurs%20kulturolozi/philosophy.gif.gif-kuhn-thomas-mirror-02.jpg

Fig. 28.

T. Kuhn (1922-1996) (Figure 28) in his seminal book, *The structure of scientific revolutions,* described historical processes and mechanisms which govern scientific practice and the development of Science [24].

Kuhn proposed that scientists at any particular time share a set of background assumptions, techniques, methodologies, terminology and world-view. This set is called a **paradigm**. Paradigms have practical value, because they inspire confidence in scientists that their work is meaningful and provide guidelines concerning data acquisition, interpretation, experiment design and so on. Paradigms also represent conceptual authority, a safeguard against pathological science. Most scientists work during the periods of 'normal science' when gaps in knowledge are filled in, puzzles (scientific problems) are generated and solved. All of this happens within the prevailing paradigm.

During this period there may appear a number of unsolved puzzles, anomalous data, observations and lose ends which do not fit into the existing paradigm i.e. which cannot be accommodated or explained by it.

Initially the scientific community tries to accommodate/rationalize the anomalies within the existing paradigm. When the number of anomalies becomes too large, a crisis occurs which induces the replacement of the existing paradigm with the new one; this is 'scientific revolution' or ‘paradigm shift’. The new paradigm is capable of accommodating all the anomalies and explains all of the previously accumulated knowledge.

During the time of scientific revolution the new and old paradigm co-exist uneasily for some time and it is then not straightforward to determine the validity of a theory on the basis of empirical data.

A paradigm shift implies the change in scientist’s perception of reality or world-view so that scientists using different paradigms view the same set of empirical results differently and use different languages to express it (e.g. phlogiston vs. oxygen theories of combustion; classical vs. quantum mechanics). The well known optical illusion (Figure 29) which shows the profiles of two women (young and old) can be used to represent the new and old paradigm during the time of scientific revolution. Both paradigms cover the same set of data, but they organize data in very different ways.

source: http://jiggette.xanga.com/

Fig. 29.

The new paradigm eventually gets established; it prevails by consensus of the scientific community which becomes convinced on the basis of evidence that the new paradigm is better. The paradigm choice is not dictated from the outside e.g. by political, economic or social factors as in other fields of human activity. The new paradigm ushers into another period of 'normal science' with vigorous progress and ordered patterns of activity. The driving forces behind scientific revolution are many. Empirical factors play important role because there are fewer anomalies and falsified statements within the new paradigm. However, non-empirical factors (influence of individual scientific groups, simplicity and elegance of the new paradigm, social and scientific standing of the proponents of theories, 'propaganda' for the new theory, government science policy) may also play a role. In the end the new paradigm prevails, replaces the old one and becomes established until the next 'scientific revolution' comes along (see concept map below).

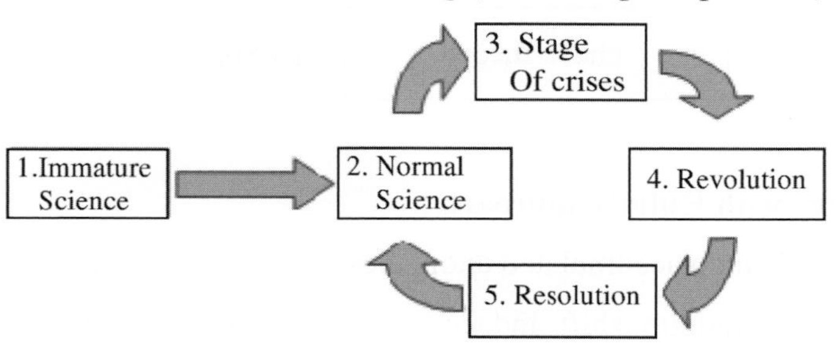

Kuhn's work has been criticized, because it suggests that there are no standards in Science and that the majority rule applies. His work influenced the philosophical description called antirealism. Here are some examples of paradigms in different fields of Science:

- Quantum mechanics (physics); see the concept map below
- Periodic table of elements (chemistry)
- Theory of evolution (biology)
- 'Central dogma of molecular biology' (life sciences)

DNA → transcription → mRNA → translation → PROTEIN

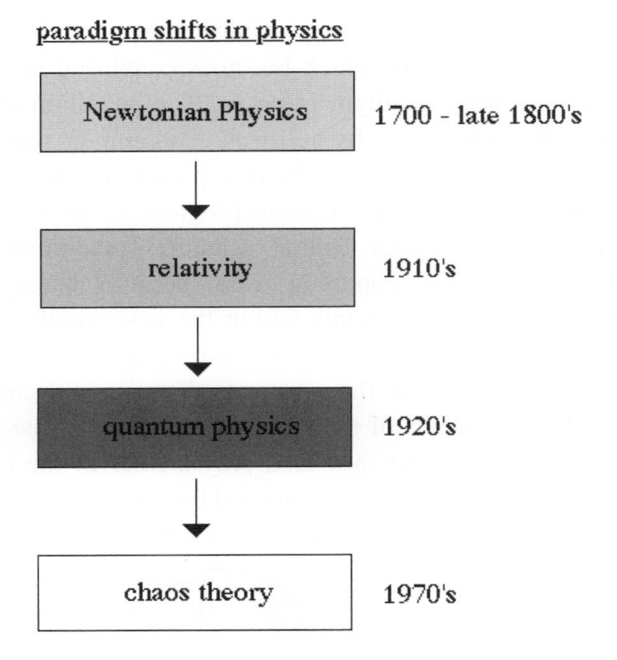

Problems with Kuhn's approach

- When do accumulated anomalies induce a paradigm shift?
- If paradigm shift induces new perception of the World and scientists from different paradigms have difficulty in communicating, how can Science progress?

The important question is when can the observation/discovery be considered truly anomalous? This brings us to the notions of revolutionary and pathological Science (see the concept map below for explanation where pathological Science fits in the Science cycle). Pathological Science involves serious investigations along the lines of inquiry which are eventually proven erroneous. Pathological Science is different from pseudo-science or fraud, which are deliberate distortions of scientific ideas and practice!

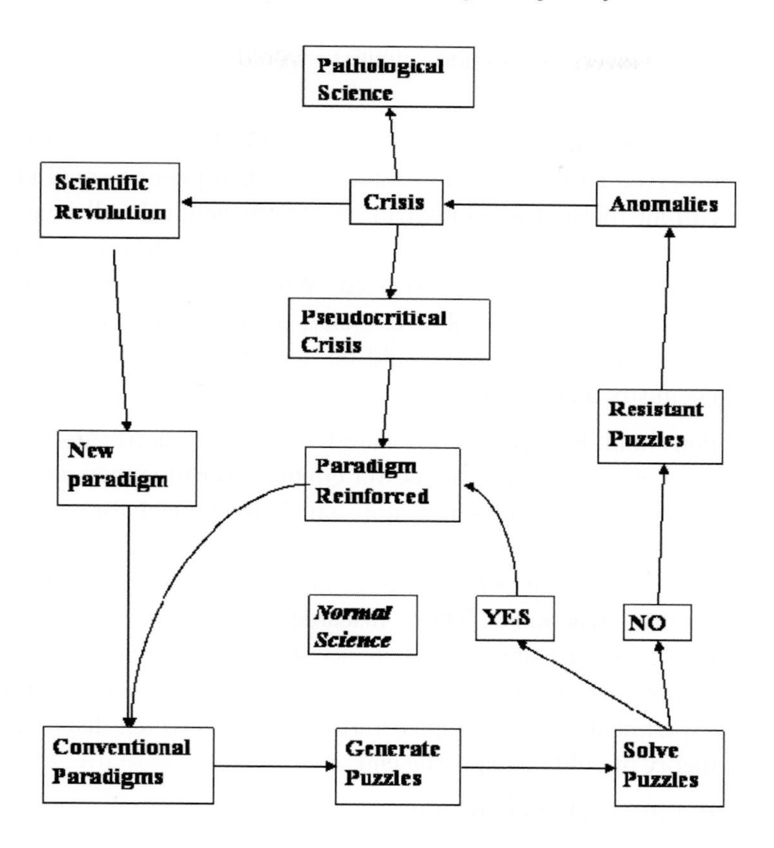

Examples of pathological science:

- **Polywater:** In 1960s, Deryagin and coworkers produced a dense liquid by condensing water in small capillaries and claimed that this liquid represented new polymerised form of water. The observations were reproducible and were made using proper scientific methods with control of experimental variables. However, 'polywater' turned out to be an artefact due to minute impurities present in ordinary water. When better purification methods and tests demonstrated the source of anomaly, Deryagin readily and honestly retracted his original conclusion. This is in accordance with ethical practice of Science where evidence takes precedence. A pseudo-scientist would never make such a retraction.

- **Cold fusion:** In 1989, Stanley Pons and Martin Fleischman have observed anomalous temperature rise when passing small electric current through a solution in an electrochemical cell containing heavy water and palladium catalyst. The palladium has absorbed hydrogen, as could be expected. What was unexpected and has remained controversial ever since, is the reported creation of excess heat; more than could be accounted for by conventional chemistry and physics.

  Pons and Fleischman measured this heat calorimetrically and when they calculated the power emitted over time they got energy of millions of joules per mol, which is atypical for a chemical process. They explained their observation by saying that the temperature rise, which was genuine, could only be due to the nuclear reactions. They discarded alternative explanations. However, nuclear reactions are accompanied by neutron emissions and other forms of radioactivity which were not detected in these experiments. This prompted further research which disproved the theory of cold fusion in the end.

- **Paranormal phenomena:**

  These phenomena are widely publicized, but have often been proven to be due either to fraud or have rendered irreproducible experimental results. Since the reproducibility of experimental results is one of the cornerstones of scientific method, this casts doubt on the results and validity of 'paranormal science'.

Common characteristics of pathological science [25]

- Maximum observed effect is produced by an agent of barely detectable intensity. The magnitude of the effect is independent of the magnitude of the cause.

- The effect has magnitude close to the detection limit of the equipment and many measurements are required because of low statistical significance of the result
- A great accuracy is claimed in the experiment
- Unusual theories, contrary to established knowledge or paradigm are suggested as explanations
- There is a failure to consider possible alternative explanation within the existing paradigm which can account for the 'extraordinary' result
- Criticisms of the results are met by 'ad hoc' explanations and are poorly thought out
- The ratio of critics/supporters of pathological results initially rises up to 50% and then gradually decays
- The remarkable result applies to a 'special system'
- Some special technique or equipment is claimed to have been used in the research
- The results require drastic revision of the existing paradigm which however fully explains results in comparable systems including those studied by the authors of pathological results themselves

The causes of pathological Science and how to avoid it?

Causes can be due to poor scientific method and to the flaws of human nature.

Discovery of a true anomaly leads to great scientific prestige, because few anomalies ever lead to revolutionary changes in Science. Media attention, professional standing, potential Nobel prizes, pressure from outside the scientific community (funding agencies) may all encourage self-delusion and extraordinary claims. The existing paradigm may be regarded as a 'prison' preventing researchers from following new ideas, but it is also a conceptual authority which prevents wild speculations, erroneous ideas and results from entering Science and wasting scientists' time in checking them out.

Ideas/results which are 'very surprising' and lie outside the reigning paradigm have high novelty and intellectual potential. High surprise investigations involve both high-gain and high-risk and require skepticism to balance it. Scientists normally work so that oddities are disproved and the existing paradigm reinforced. The greater the departure of their results from the established paradigm, the more aware scientists must be of possible mechanisms by which cognitive errors may occur. Nevertheless, scientific research would be impossible if it was always performed safely within the established paradigm!

Safeguards against pathological results:

- Consider and test several plausible hypotheses to explain the extraordinary results.
- Use the prevailing paradigm as a guide until one becomes certain that a revision of paradigm is needed.
- Ensure that results are reproducible.
- Use statistical arguments (statistical significance, error intervals) conservatively.
- Test the observed extraordinary phenomenon by several independent methods.
- Discuss the surprising findings with peers (formally and informally) and consider their critiques very carefully.
- Try hard to falsify your experiment/theory/interpretation.
- Acknowledge due falsification of your hypothesis; such communal corrective process is essential for Science and makes it work.

Note: The key principle is that 'extraordinary claims require extraordinary evidence'.

Sometimes it is not easy to distinguish between fraud and pathological Science as example in Ref. [26] about the production of artificial diamonds demonstrates.

2. Jigsaw puzzle

Scientific activity can also be likened to assembling the puzzle game from the given pieces [8]. Pieces of the puzzle are given to each participant. Each participant then tries to fit a piece in, while his actions are observed and corrected by other participants. Other participants then try to work out how the next piece is to be fitted in etc. (see the cartoon below).

source: http://www.utoronto.ca/jpolanyi/public_affairs/5a.gif

The puzzle game model is very useful because it helps us to understand and "visualize" many characteristics of modern Science in a straightforward way. It explains how modern Science arises out of systematic cooperation between scientists, how Science works best without outside constraints e.g. ideology, why pseudo-scientists working in isolation from the scientific community produce wrong results. Let us compare the "scientific puzzle" with the puzzle game variety. Everyday experience tells us that the difficulty in assembling the puzzle is proportional to the number of pieces present (provided the pieces do not have identical or regular shapes). In Science the number of pieces is very, very large. Furthermore, we are not even sure whether we have all the pieces needed to assemble the puzzle (e.g. some new as yet undiscovered facts). We also do not know what the finished puzzle looks like, unlike in the puzzle game where the final template is given. This uncertainty is what attracts scientists to Science, but it also explains why Science is a difficult activity. We cannot prove scientific theories in an absolute sense because we do not know what the final form of the puzzle template (if there is one) looks like. However, scientific results are not arbitrary social constructs because the puzzle pieces are matched together. They cannot be assembled in an arbitrary way! In other words the scientific information (data, theories, laws, hypotheses) must be logically consistent! The consistency is no guarantee of Truth since new evidence (puzzle pieces) may emerge in the future, but it is nonetheless a strong indication that we are on the right track. This is important when considering the objections against Science raised by Postmodernists.

3. Knowledge filter

The filter model [8] (Figure 30) shows that:

- scientific knowledge arises from large and diverse 'cocktail' of claims, results, ideas by gradual filtering. Filtering process is done by the members of scientific community who are guided by experiment.
- 'frontier science' is not completely reliable. It is just the latest information that has been made widely available through scientific publishing.
- The most reliable form of scientific knowledge is 'textbook science'. It is impersonal and it is what students learn. Because of reliability and impersonality of 'textbook science' people get the impression that Science uses well defined methods and procedures ('myth of a scientific method'). 'Textbook science' lags in time behind 'frontier science'. Media always report results of 'frontier science' which creates confusion and leads to misunderstandings/misconceptions about Science if such results are later retracted. This confusion between ‘frontier science’ and ‘textbook science’ leads to public unease about fraud in Science.

Frontier science may be unreliable and fraud occurs in the domain of "frontier science', but the final *filtrate* ('textbook science') is always very reliable and useful.

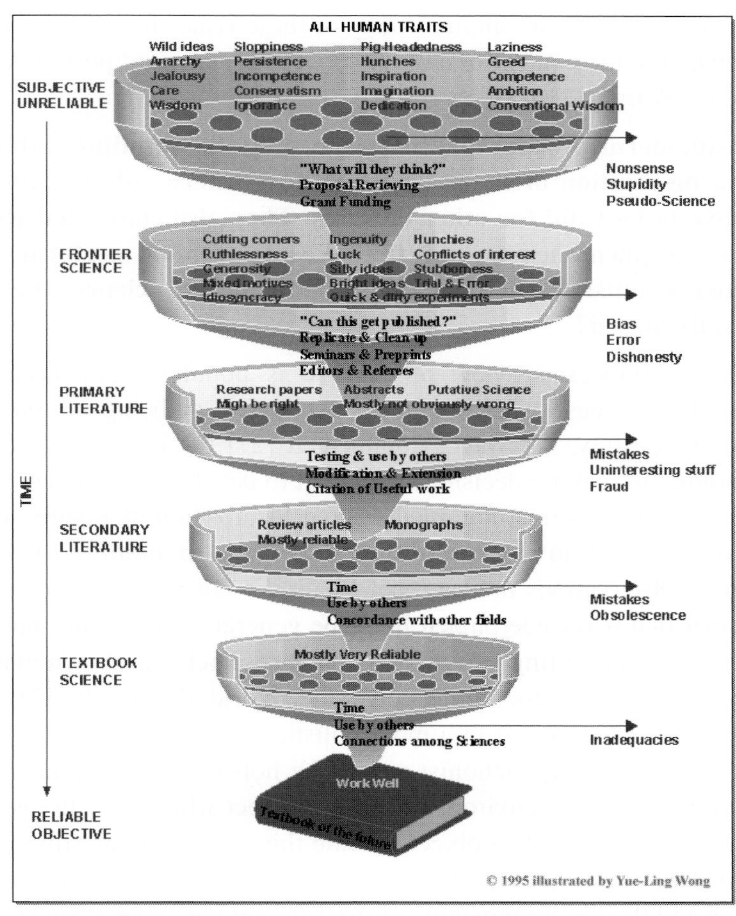

source: http://www.files.chem.vt.edu/chem-ed/ethics/hbauer/hbauer-filter.html
Reprinted with permission of H.H. Bauer

Fig. 30.

For a further discussion of these models see reference [8].

The validity of scientific results

In this section we shall discuss some questions and answers regarding Science. The purpose is to describe alternative points of view and give pointers to assess their validity. Many of these issues have been and still are debated in the context of a clash between realists and postmodernists ("Science Wars") [21d].

Q. Postmodernists claim that Science is a Western tradition; that it is a good description of the World for the Western culture, but does not have to be valid for other cultures. They also claim that Reality is a local phenomenon and that Reason cannot perform value judgments outside such local phenomena. Is Science therefore culturally biased?

A. The postmodernist view of Science is popular today because it appeals to the sense of freedom and the notion of being in control of one's own destiny. As Mayor commented [27] the use of Science and Technology in making decisions may lead to devaluing of human factor. However, postmodernist arguments about localization of Reality lead to *relativism* (belief in whatever one wishes) or to *nihilism* (belief in nothing). Observations indicate that Reality is not a local phenomenon, but rather that it is underpinned by some general, widely valid laws and patterns. It is interesting to note that recent research has shown that East Asians and Westerners do perceive the World differently [28]. The former group perceives it in a more holistic light, while the latter group tends to have more reductionist view. The holistic view emphasizes the background (context, environment) of an object while reductionist view emphasizes the individual object. While this result is interesting it does not invalidate scientific results. The complexity of interactions between humans and their environment leads to different outcomes ("views of the World"), but it does not prove that these interactions cannot be described by scientific laws and processes. After all, the researchers who published this article [28] used scientific laws and principles in arriving at their conclusion!

Q. Scientists claim that Science is self-sufficient, that it needs no justification/grounding, that it is the sole source of knowledge. Is this true?

A. This claim of self-sufficiency sounds like "begging the question" fallacy. The practical success of Science has tempted many scientists and non-scientists to consider Science as the ultimate form of knowledge! Yet the claim that Science is the sole arbiter of Reality and can explain everything is itself a claim made outside Science (a metaphysical claim, 'scientism'). The problem is that Reality becomes defined in terms of what can be known through Science. Science evolves, so which Science should be used as the arbiter of knowledge: present or future as yet unknown Science? Should Reality be defined in terms of human ability to discover it?

Claims based on 'scientism' leave unanswered questions like: Why is the World ordered and why is human mind capable of understanding the World?

A possible answer to this question may go like this. "Our minds have evolved through natural selection to understand the World in order to enhance our chances of survival." This sounds plausible. However, it also imputes purpose to Evolutionary process which certainly is not true. The development of our mental faculties and reasoning ability may have been the consequence of excessively efficient human adaptation to environmental pressures! But why it happened only in humans? There is not a conclusive, verifiable answer at the moment so speculation on the issue shall continue. Trigg [29] has suggested that Science needs to have a rational foundation outside its own domain and that blind faith (acceptance of authority without questioning) in Science can be as harmful as blind faith in a religious doctrine. However, the problem with Trigg's suggestion is that it does not provide any guidelines about how this external foundation can be found and its suitability assessed.

Q. Is Science is only a cultural artifact? Are the results of Science just social constructions?

A. Science is the product of culture, a historical phenomenon and cannot be justified as the absolutely superior mode of thinking. However, this does not imply that all modes of thinking are equally valid or that Science is purely arbitrary construct of human mind and therefore does not represent Reality. The challenge to Science from Postmodernist position is related to the worry about "devaluing the human factor" which can happen if we allow Science to dominate decisions in our lives [27]. Instead, various ethical, humanist principles of common good and tolerance are suggested as an antidote to all-encompassing natural laws. "Devaluing human factor" is however not evidence that can be used in respect of the validity of scientific claims. It is an 'appeal to emotion' type fallacy. This comment does not of course suggest that 'human factor' is unimportant; it only urges that sound rational arguments about the role of Science be put forward. Science is a rational enterprise related to Nature which can apparently be understood in a rational way. Emotions have their important role in other fields of endeavor, but not in the criticisms of scientific results. In Chapter 3 we shall discuss further this sensitive issue and controversies about human ethics and Science.

While Science is not purely a social construct there can be no "proof" of Reality engendered by Science. As physicist Planck has said:

> *"Science cannot solve the ultimate mystery of Nature. And that is because in the last analysis we ourselves are part of nature and therefore part of the mystery that we are trying to solve".*
>
> *M. Planck*

Leaving the issue of the proof of Reality aside, one must add a note of caution. The ethical comments about 'human factor' are anthropocentric and depend on the validity of certain assumptions. One such assumption is that human civilization **must** survive and keep increasing its material prosperity. While this is a necessary principle for our Societies to function it is also a logical fallacy ("appeal to emotion").

Q. Quine-Duhem thesis shows the under-determination of scientific theories by empirical data. Are theories then arbitrary constructions?

A. It is not easy to come up with alternative explanations for a multitude of data and at the same time have all the explanations match up coherently.

Furthermore, scientific explanations and theories often involve predictions of unknown and unexpected phenomena and discoveries. How can one put forward a Theory which correctly predicts something which was unknown when the Theory was forged? The example (Figure 31) shows how Theory of Relativity proposed by Einstein predicted the bending of light rays from distant stars when they pass through Sun's gravitational field. This was something totally unexpected and was not used in the design of the Theory of Relativity! The ability of scientific theories and ideas to go beyond the boundaries of knowledge from which they were constructed, strongly suggests that in some way scientific knowledge does reflect Reality and is therefore not arbitrary design. This does not entail that scientific knowledge is complete, that it is the sole representation of Reality or valid for all possible levels of Reality. When scientific theories go beyond the boundaries of current knowledge it is crucial that they are falsifiable as Einstein's Theory of Relativity surely was. Detection of the bending of light beam was certainly possible and when it was detected it confirmed his Theory.

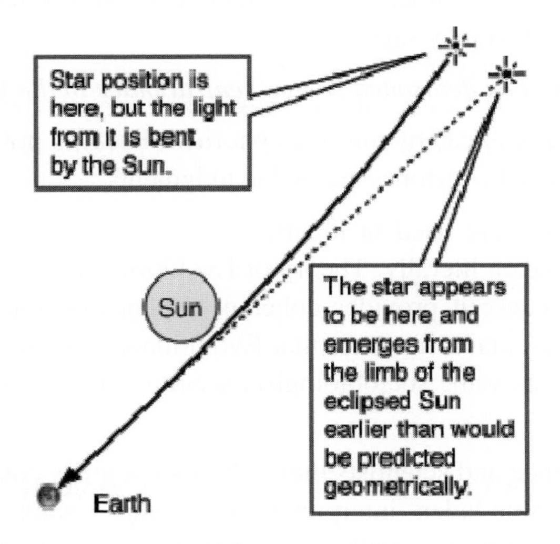

source:http://hyperphysics.phyastr.gsu.edu/HBASE/relativ/imgrel/dstar.gif

Fig. 31.

However, some current scientific theories are not testable. One example is the cosmological theory proposing the existence of parallel, timeless Universes (multiverses) [30]. Such cosmological theories, while imaginative and exciting are not amenable to experimental verification and hence should be taken with a pinch of salt [30]. Nonetheless, this ability to transcend the present state of scientific knowledge is an important driving force in Science as long as ideas and concepts are subject to continuous experimental observations and reevaluations. Let us remember that Aristarch's theory of heliocentric system (see Chapter 4) was initially unverifiable and speculative due to the primitive state of experimental technology in his time. However, his theory was in principle falsifiable and much later it was confirmed by observations.

Caveats

The interpretations of scientific results must be made carefully. Many criticisms of Science have to do with questionable, metaphorical interpretations of scientific results rather than with Science itself. For example, a scientist may say:

'Humanity has been dethroned from a central place in the Universe.'

Such a statement is catchy and metaphorical, but it is not scientific and contains emotional overtones and value judgments.

While metaphors are used in theories as an aid to understanding, they should not be taken literally. Theory of Evolution has a well defined role in Science, because it provides coherent explanation for a large set of observations. To claim however that Evolutionary Theory should be the basis of ethics as some sociobiologists seem to imply is not a scientific claim.

Critics of Science and scientists themselves sometimes confuse scientific results with value-laden interpretations of these same results. Such interpretations are non-scientific even if they are proposed by scientists! An example of such value-laden extrapolation from scientific data and a commentary on such practice is given in the quotations below.

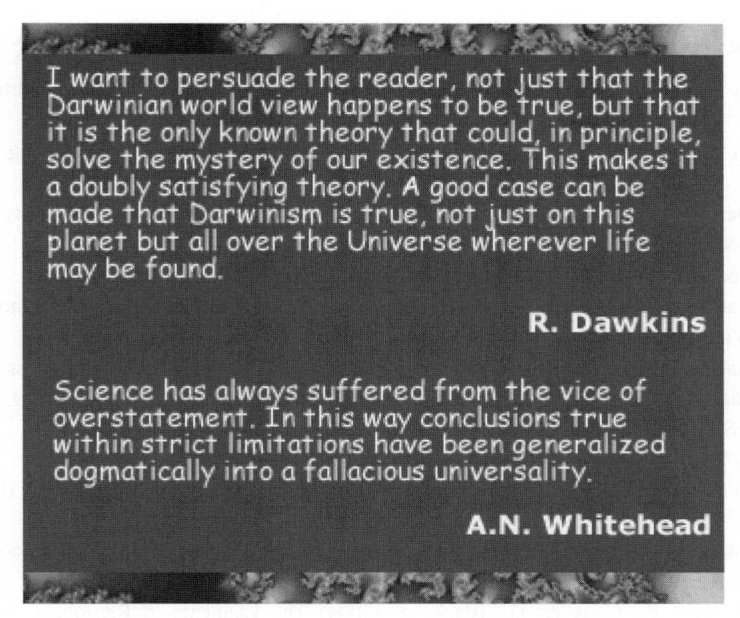

I want to persuade the reader, not just that the Darwinian world view happens to be true, but that it is the only known theory that could, in principle, solve the mystery of our existence. This makes it a doubly satisfying theory. A good case can be made that Darwinism is true, not just on this planet but all over the Universe wherever life may be found.
R. Dawkins
Science has always suffered from the vice of overstatement. In this way conclusions true within strict limitations have been generalized dogmatically into a fallacious universality.
A.N. Whitehead

Of course, genes and Evolution have determined our identity, but they are not the only determining factors nor should one derive moral norms from empirical observations ('naturalistic fallacy'). Likewise, political ideologies, economic interests and cultural prejudices certainly have some influence on the direction of scientific research, but they are not important factors which determine its final outcome.

Science has Nature as its main and independent reference point and this precludes scientific research from becoming governed purely by social or cultural agendas, as postmodernists wrongly claim.

Reflection:

What is the main reason for having confidence in scientific methods and results?

Tips: consider how the scientific results are arrived at and how are they verified. Recall the jigsaw puzzle model of Science.

References for Chapter 2:

1. a) Mackay, A.L. (1991) *A Dictionary of Scientific Quotations*, (IOP Publishing, UK).
   b) Gaither, C.C., Cavazos-Gaither, A.E. (2000) Scientifically Speaking, (IOP Publishing, UK).
2. Derry, G.N. (1999) What Science is and How it Works, (Princeton University Press, USA).
3. Lee, J.A. (2000) The Scientific Endeavour, (Addison-Wesley Longman, USA).
4. Carey, S.S. (2004) A Beginner's Guide to Scientific Method, 3rd ed., (Thomson-Wadsworth, USA).
5. Gauch, H.G. (2003) Scientific Method in Practice, (Cambridge University Press, UK).
6. Giere, R.N. (1997) Understanding Scientific Reasoning, 4th ed., (Harcourt Brace, USA).
7. a) Thompson, M. (2001) Philosophy of Science, (Hodder & Stoughton, UK).
   b) Okasha, S. (2002) Philosophy of Science: A Very Short Introduction, (Oxford University Press, UK).
8. Bauer, H.H. (1992) Scientific Literacy and the Myth of the Scientific Method, (The University of Illinois Press, Chicago, USA).
9. Munson, R., Conway, D., Black, A. (2004) The Elements of Reasoning, 4th ed., (Thomson-Wadsworth, USA).
10. Hoffmann, R., Minkin, V.I., Carpenter, B.K. (1996). Ockham's razor and Chemistry, Bull.Soc.Chim.Fr. 133, pp.117-121.
11. Gott, R., Duggan, S. (2003) Understanding and Using Scientific Evidence, (Sage Publications, UK).
12. Richards, J.R. (2000) Human Nature after Darwin, (Routledge, UK).
13. a) Dawkins, R. (2006) The God Delusion, (Bantam Books, UK).
    b) McGrath, A. (2007) The Dawkins Delusion, (SPCK, UK).
    c) Ruse, M., (2009) Dawkins et al bring us into disrepute, Guardian Newspaper, (2 Nov, UK).
    d) Alexander, D. (2009) Intelligent design is not science, Guardian Newspaper, (Dec 3, UK)
14. Hung, E. (1997) The Nature of Science: Problems and Perspectives, (Wadsworth, USA).
15. Laughlin, R.B., Pines, D. (2000). The theory of everything, Proc.Natl.Acad.Sci.USA, 97, pp.28-31.
16. Wolfenstein, L. (2003) Lessons from Kepler and the theory of everything, Proc.Natl.Acad.Sci.USA, 100, pp.5001-5003.
17. Kutzelnigg, W. (2000). Perspective on "Quantum mechanics of many-electron systems"', Theor.Chem.Acc. 103, pp.182-186.

18. Schuster, P. (2007). A beginning of the end of the holism versus reductionism debate?: Molecular biology goes cellular and organismic, Complexity, 13, (Sept/Oct issue) pp.10-13.
19. Abbott, R. (2009). The reductionist blind spot, Complexity, 14, (May/Jun issue) pp.10-22.
20. Kurzynski, M. (2006) The Thermodynamic Machinery of Life, (Springer Verlag, Germany).
21. a) Snow, C.P. (1993) The Two Cultures, (Cambridge University Press, UK).
    b) Crease, R.P. (2009). 'Two cultures' turns 50, Phys.World, (May issue), pp.19.
    c) Waugh, M. (2009). Last retort: Looking through the Snow, Chem.World, (Dec issue), pp.88.
    d) Appignanesi, E. (2002) Postmodernism and Big Science, (Icon Books, UK).
22. Berson, J.A. (2003) Chemical Discovery and the Logicians' Program, (Wiley-VCH, Germany).
23. Buskirk, A., Baradaran, H. (2009). Can reaction mechanisms be proven?, J.Chem.Educ. 86, pp.551-558.
24. Kuhn, T.S. (1996) The Structure of Scientific Revolutions, 3rd ed., (The University of Chicago Press, USA).
25. Turro, N.J. (2000). Paradigms Lost and Paradigms Found: Examples of Science Extraordinary and Science Pathological - And How To Tell the Difference, Angew.Chem.Int.Ed.Engl. 39, pp.2255-2259.
26. Gölitz, P. (2004). Editorial Note: Diamond Synthesis in Doubt, Angew.Chem.Int.Ed.Engl. 43, pp.4687.
27. Mayor, F. (1998). Devaluing the human factor, The Times Higher Education Supplement, (Feb 6, UK) pp.13.
28. Nisbett, R.E., Masuda, T. (2003). Culture and point of view, Proc.Nat.Acad.Sci.USA 100, pp.11163-11170.
29. Trigg, R. (2002). Does science deal with the real world?, Interdiscip.Sci.Rev. 27, pp.94-99.
30. Smolin, L. (2009). The unique Universe, Phys.World, (June issue) pp.21-26.

Chapter 3

Sociological and Ethical Aspects of Science

(citations in this Chapter were taken from refs. [1] and [20]).

Some fundamental questions considered in this Chapter:

How is Science perceived by the public and why?
Who determines how scientific results are used?
What are the overall benefits/costs of scientific progress?
Can moral norms be scientifically based?
How do Science & Society interact?
What is the relationship between Science & values?

Science and Society today interact strongly through various channels, because Science is a human activity essential for the existence and progress of modern civilization (see the concept map below). Like in all human activities questions arise about goals and values and how can they be realized.

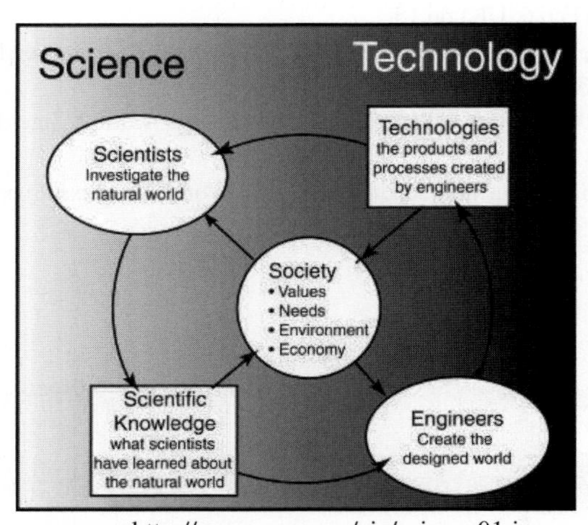

source: http://www.mos.org/eie/science01.jpg

The sources of friction between Science and public arise because of the character of Science itself and because of misunderstanding of what Science can (or cannot) do. There is also a wide range of other misconceptions amongst general public regarding Science [4,5]. Here are some examples of misconceptions/concerns about Science with relevant comments:

- *Science is dangerous because scientists are arrogant people with unlimited knowledge and power; there is no public control over Science*

Scientists are a social group with specialized knowledge (like lawyers, doctors, car-mechanics) and are self-regulating perhaps more than any other profession. The self-regulation of Science involves international scientific community while other professions have national regulating professional bodies.

The boundaries and activities of Science are influenced from within and from without. From within: Science has complex infrastructure (scientific meetings, peer-review, funding and publications, professional bodies, learned societies). Because Science is often driven by curiosity, it is unregulated and hence may seem 'dangerous'. However without curiosity there can be no progress in Science.

Governing social elites (political, business) have little educational background in Science yet they control Science from without by regulating its funding and influencing research programs. So much for the 'unlimited knowledge and power' (see the quotation below)!

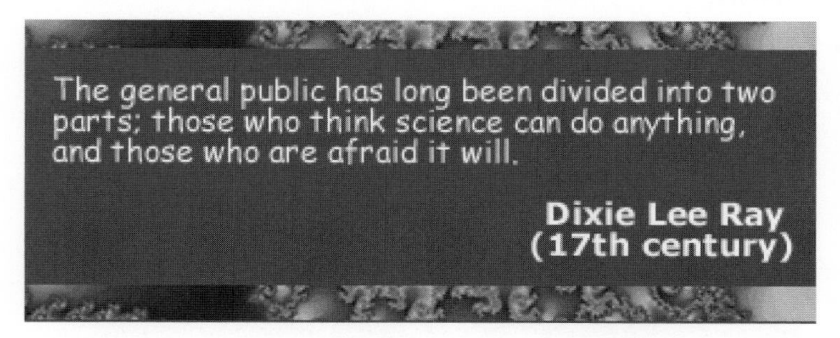

The general public has long been divided into two parts; those who think science can do anything, and those who are afraid it will.
Dixie Lee Ray
(17th century)

- *Scientists are irresponsible and unethical because they claim that their results are value-neutral/value-free.*
- *Science is unrelated to real life concerns; it is alienating and boring (see quotations below).*

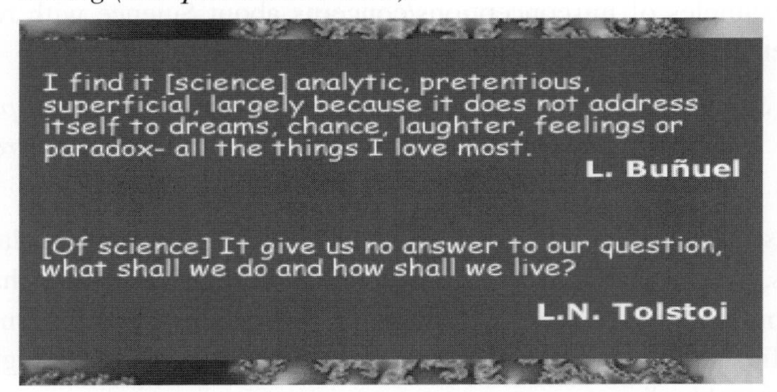

I find it [science] analytic, pretentious, superficial, largely because it does not address itself to dreams, chance, laughter, feelings or paradox- all the things I love most.

[Of science] It give us no answer to our question, what shall we do and how shall we live?

Science is the best available method for exploring observer-independent facts about Nature, but that does not make scientific results value-free. This is because Science rests on "two pillars": laws of Nature and social relationships (Science is a human activity). However successful Science might be, it is not designed to answer questions about values, ethics, beauty, human relationships, meaning of Life.

- *Scientists claim that Science is the privileged form of knowledge which has the power to explain all of Nature ('scientism').*

Scientism is not an empirically justified scientific theory, but a controversial philosophical hypothesis/extrapolation based on scientific achievements.

- *Science is difficult to understand, it is obscure.*

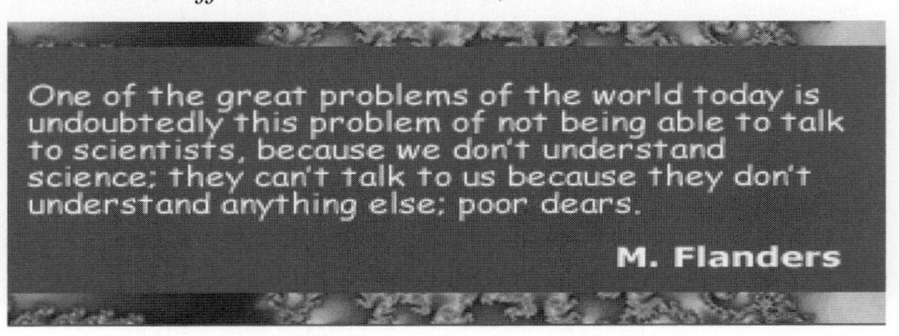

One of the great problems of the world today is undoubtedly this problem of not being able to talk to scientists, because we don't understand science; they can't talk to us because they don't understand anything else; poor dears.
M. Flanders

Science uses jargon and mathematical code which is difficult to comprehend by general public, especially in view of the current lack of emphasis on mathematical and scientific training in compulsory education. However, this code is used because it is very efficient and precise. Science investigates processes which may not be directly observable or are counterintuitive to our mind (e.g. quantum mechanics) and hence cannot rely on expressiveness of human language alone. Even though human language is very subtle and sophisticated, human language often lacks precision and unambiguous meaning which are required in the study of Nature. It is also true that scientists are often poor communicators due to difficult training and demanding nature of their work (see quotation above). This leaves them little time or energy for acquiring communication skills.

Two visions of science

Two extreme perceptions of Science which have emerged are called [4]:

- Baconian vision (after F. Bacon)
- Frankensteinian vision (after the novel *Frankenstein*)

Baconian vision:

This vision rests on the belief in unlimited progress of Mankind which can be achieved through the application of Science. English thinker F. Bacon (1561-1621) was the first to suggest that great benefits for Society may be obtained through systematic scientific research and application of scientific knowledge in the exploitation of natural resources.
The photo in Figure 32 which shows the advertisement for HSBC bank expresses this perception admirably well. Knowledge is power! Since knowledge is firmly taken by general public and governments to be a guarantor of 'success' and 'happiness' in life, children are encouraged to compete and excel in the acquisition of knowledge. A large portion of this knowledge comprises scientific and technological knowledge. It is not surprising therefore that some of the most ardent proponents of this vision can be found in the developing countries.

source: http://hankwhittemore.files.wordpress.com/2009/07/francis_bacon-rosicrucian.jpg / source: author's personal collection

Fig. 32.

The Baconian vision originated during the age of 'Enlightenment' (in 17-18th century W. Europe). Enlightenment was an intellectual movement which believed that human conditions (physical and social environment) can be controlled and improved mostly through Science, Technology and human Reason. This is a uniquely European idea which was not discernible in other civilizations. This idea is even today the driving force behind most developments in the World. When put into practice, the idea lead to the pre-eminence of Western societies up to the 21th century. With the rise of globalization the idea has now been adopted worldwide. All countries and their governments strive to use Science and Technology for improvement of lives of their populations irrespective of political, social or religious heritage of their societies. Like any far-reaching social or religious idea it was and is prone to abuses (e.g. the use of advanced technology for colonial conquest) even though it also provided great universal benefits e.g. improvement in health-care, life expectancy, communications etc.

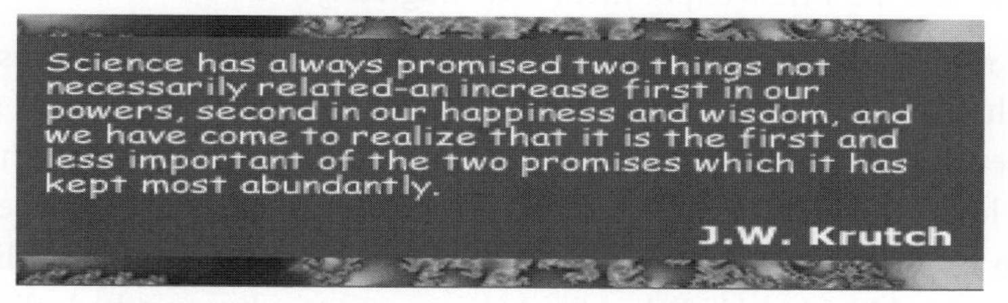

Science has always promised two things not necessarily related-an increase first in our powers, second in our happiness and wisdom, and we have come to realize that it is the first and less important of the two promises which it has kept most abundantly.
J.W. Krutch

Frankensteinian vision:

source: ttp://www.wpclipart.com/famous/writer/writers_M_to_S/Mary_Shelley.png.html

Fig. 33.

This vision is named after the novel by M. Shelley (1797-1851) and embodies scepticism and pessimism about human progress through ST developments, especially in view of the role of ST in weapons developments, environmental degradation, genetic manipulations and ST support for unrestricted economic development. This scepticism (being most prevalent today in developed Western countries) started in the early 19th century Europe [5] and expressed itself through various cultural movements e.g. romanticism.

The novel *Frankenstein or the Modern Prometheus* was the first modern science fiction novel and one of the first novels from the era of Romanticism which followed the Age of Enlightenment. Frankenstein is a doctor who creates a gigantic living man from the parts of corpses he finds in tombs. The monster is rejected by society because of his grotesque looks, becomes unhappy about his rejection, he starts killing people and the whole experiment ends tragically.

This novel is interesting for its uncanny premonition of modern research into human cloning and genetic engineering, but it also shows that social concerns regarding dangers of tampering with natural processes have a long history, predating modern times. A good example of such concerns is given by Silver [6]. He described a possible future scenario according to which the Mankind shall be divided into genetically enriched elite and educationally impoverished masses which cannot breed successfully with their gene-enhanced betters [6]. This is indeed a chilling prospect whose probability may be reduced (somewhat?) by the fact that genes in themselves have less influence on human lives than previously thought [7] and because other factors (e.g. social environment) are also as important. The ever present mutation processes may also scramble genetic superiority. Finally, let us remember that all systems in Nature are strongly interconnected and continuously evolving. The attempts to produce genetically engineered 'super race' would likely result in generating unsustainable gradients within human population itself and between human population and the Environment because of excessive energy consumption required to maintain these gradients. 'Nature abhors gradients' and works efficiently to degrade them [8]. This is probably the main reason why such 'superhuman' possibilities belong to the realm of Hollywood imagination.

Other important novels written in a similar vein are: A. Huxley, '*Brave New World*' and G. Orwell, '*1984*'. Many ancient myths also express similar concerns through metaphors and allegories:

- *Prometheus and Pandora* (Greek mythology) - Pandora opens her gift-box out of curiosity and all evils are let loose on the World. Prometheus gives fire to Man and is punished by Gods by having his liver devoured by birds (Figure 34).

sources: http://www.wpclipart.com/world_history/lore/pandora.png.html
http://www.wpclipart.com/world_history/lore/prometheus.png.html

Fig. 34.

- *Biblical Creation story* - Adam, Eve, Garden of Eden - Man and Woman gain knowledge and power by eating fruit from the Tree of Knowing of Good and Evil, they lose innocence and are expelled from the Garden.

source: http://www.churchtimes.co.uk/uploads/images/
The%20Garden%20of%20Eden%20and%20the%20Fall%20of%20Man%231%23.jpg

Fig. 35.

Modern societies often consider these ancient concerns to be out-of-date and fanciful and rely on ST to pull us through any problems. Such reasoning is simplistic. It is caused by misconceptions about ST and the fact that market-driven media are the preferred source of information about ST for a majority of people. Markets can be influenced by fancies, human expectations and emotions rather than sound logical judgements.

Scientific interpretation of these old concerns is that many complex systems (e.g. human societies, countries and our global Environment) are open thermodynamic systems whose behaviour is governed by energy and mass transfers, they are organized on the principles of thermodynamics and subject to boundary conditions [8]. Boundary conditions are related to properties and states of other systems which interact with the system under consideration; they are related to our Environment. Open systems are those where energy and matter is continuously transferred across the system's boundaries. This openness prevents the system from ever reaching an equilibrium state. Temporary stability only can be achieved at the expense of energy intake. The interactions between open systems also suggest that perturbation will readily propagate from one system to the next.

A complementary scientific description of Society as a network of linked individuals has been proposed by Csermely [8b]. Csermely claimed that weak links between individuals stabilize societies and has tried to explain historical changes which societies undergo as a consequence of redistribution of links within a particular society (network transitions). His model is interesting, but is topological and descriptive. It describes various types of possible social networks and some of their properties, but it does not explain dynamical processes which drive transformations of one type of network into another. In order to understand the dynamics of social change one needs to take energy input into account.

We suggest that the characteristics of open thermodynamic systems (systems being e.g. individuals, nations, societies and global environment) may preclude global society from ever achieving the stability of 'golden age' represented by peace, justice and universal well being. It is the undisputed fact that universal peace and prosperity have never yet been achieved in human history for any significant length of time. Whether this desirable state can ever be achieved in the future (as social elites in many countries insist is possible) remains to be seen. This claim sounds very pessimistic and Malthus-like, but 'pessimistic' is not a valid counterargument in a scientific debate.

How are these comments related to general concerns regarding interactions between Man and Environment?

The interactions between systems necessarily impose checks (boundary conditions) on uncontrollable growth and other excesses of individual systems. 'Tampering with Nature' as described in the myths above is possible only through the imposition of gradients which themselves are only achievable by degradation and dissipation of energy [8]. Nature has a tendency (due to statistical probabilities of various events) to remove gradients. When we create excesses (high gradients) in our Environment or Society such gradients are bound to be reduced sooner or later with often unwanted consequences. Let us look at some examples.

Examples of unstable, high gradients may be local, national, consumer 'paradises' which can only exist at the expense of high energy intake and pollution generation. Since World has no access to inexhaustible energy supply, maintenance of such gradients is impossible over a long time. The reduction of gradients via pathways which we associate with undesirable events: climate change, pollution, economic, social and political turmoil, then spontaneously ensues. The undesirable events certainly temporarily reduce gradients or our appetite/ability to create gradients. Uncontrolled economic and population growth has created global and national gradients of wealth and population density. These gradients are currently being reduced spontaneously by various undesired pathways (illegal immigration, terrorism, military conflicts). Can we find more desirable pathways? Perhaps we can, but that would require giving up some cherished notions about preferred lifestyle and aims of Life. Even if we have no desire to control such growth, the nature of dynamical systems will spontaneously enforce gradient degradation. The Environmental and population problems are thus **NOT** solely or even primarily ethical, political or economic issues, but are the fundamental consequences of the dynamics of complex systems and of our unwillingness to recognize this fact-of-life.

On a lighter note let us recall how English physicist Snow (see Chapter 1) jokingly and poignantly summarized the three Laws of Thermodynamics as:

"*You cannot win*" (1st Law)

"*You cannot break even*" (2nd Law)

"*You cannot get out of the game*" (3rd Law)

This summarizes the constraints mentioned above bearing in mind that Laws operate on groups or ensembles of entities, not on a single entity.

Reflection:

Why does general public perceive Science in an ambiguous way (as a mixture of good and bad things)?

What other human activity is perceived in a similarly ambiguous way?

Which human activity is generally perceived in an almost exclusively positive light?

Why are there two visions of Science (Baconian & Frankensteinian)?

Hints:

Consider what understanding the public has of Science.

Consider which other human belief system claims universal validity.

Consider which activity can be considered harmless

Consider which view is more difficult to hold (simple, black-and-white) or nuanced one

Is Science value-neutral?

This is a very important question, because many scientists and members of public make a distinction between investigating Nature and human manipulations of Nature (using scientific knowledge). The latter is aimed at fulfilling human goals and desires. The confusion often arises because the word Science has several connotations: scientific activity, scientific outcomes, applications of scientific results etc. The answer to the question about value-neutrality then depends on whether we are referring to scientific knowledge which is value-free or to scientific applications which are value-laden. Some commonly encountered arguments on this theme can be outlined as follows:

Proposition A: *The application of scientific knowledge is for Society to decide!*

Counter-argument: The mistake in this argument is that Society is not an entity which directly influences decisions about the applications of Science. The concept map below shows why this is so. Modern societies are like dense networks, highly structured and with many intermediaries between the mass of citizens (electorate) and the scientific research activity. The intermediaries are unelected (i.e. they are not products of democratic, but of business activities) and comprise institutions like banks, corporations etc. which are not under direct control of the Society's citizens. This has been brought to light during 2008 economic crisis when citizens had to rescue large Banks in spite of corporate malpractices. The 'social executives' who directly influence scientific research are precisely these intermediaries, especially business corporations and government organizations. Science requires considerable financial support which is provided by the intermediaries in exchange for allowing the same intermediaries to influence topics and direct scientific research.

The professionalization and industrialization of Science has also lead to some research being performed under conditions of military or industrial secrecy. This replaces the principle of Science as a free inquiry into

Nature whose results are open to all, with the principles of business practice. Concept map below describes this counter-argument.

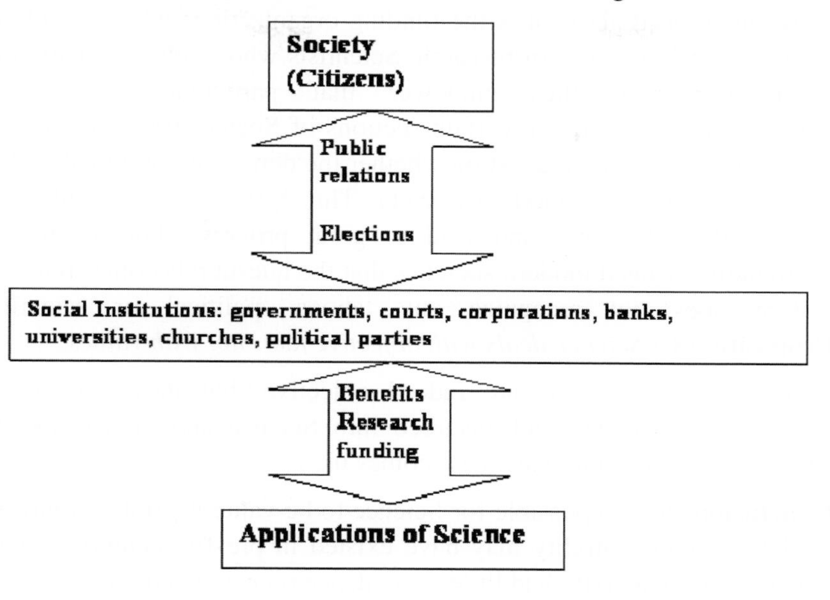

Proposition B: *The only value recognized by scientists is the knowledge for its own sake* (see quotation below).

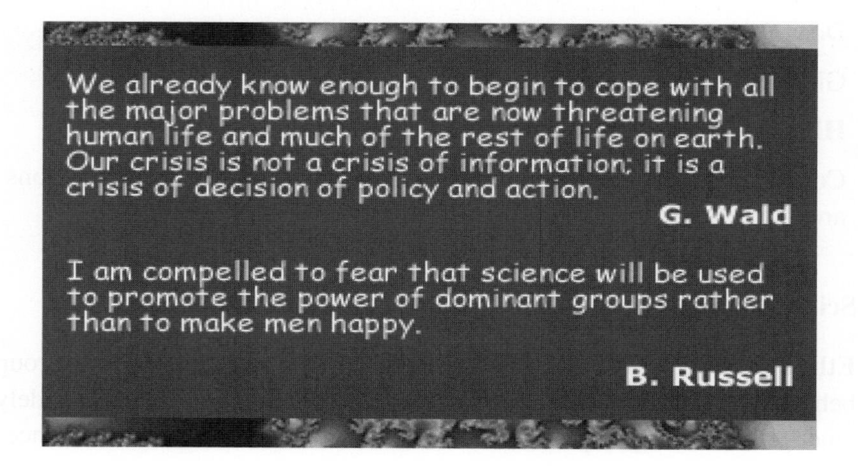

We already know enough to begin to cope with all the major problems that are now threatening human life and much of the rest of life on earth. Our crisis is not a crisis of information; it is a crisis of decision of policy and action.
G. Wald
I am compelled to fear that science will be used to promote the power of dominant groups rather than to make men happy.
B. Russell

Counter-argument: Scientists do Science for a variety of motives, not all of which have to do with the enhancement of knowledge. Furthermore, as we mentioned above it is the funding organizations which decide on the topics and priorities of research. Scientists who work for a particular organization/group thus empower that particular group. This concentration of power in certain sections of Society brought about by ST raises an interesting question whether the democratic political system is the "best one" for modern Society. This system was certainly very successful in the past and lead to great progress, but ST has so profoundly changed modern societies that the question becomes relevant. Nature does not recognize any taboos, political or scientific.
Proposition C: *Science deals with objective facts not with values.*

Counter-argument: Facts are indeed objective, but they have to be interpreted, organized and absorbed into Science and human society. This is where the value-ladenness comes in.

Conclusion: It is impossible for Science to be value-neutral in a modern Society! Value-neutrality may have existed in pre-19th century Science when scientific activity had little dependence on major funding sources.

Reflection:

Does Science enhance the development of democratic society?

Give one example in favor and one against this notion.

Hint:

Consider modern applications of Science and how these applications are implemented.

Science in Ethics

Ethics is a body of norms and rules which guide individual and group behaviour and actions in human societies. These norms vary widely amongst cultures, groups, nations, religious communities. The differences in moral norms between societies have been the historical source of major conflicts, wars etc. If we can devise universal principles of Ethics which transcend national, religious or group boundaries we may hope to reduce the extent of conflicts. What can possible universal principles be? Patzig [9] has discussed several possibilities:

- **Religion-based ethics**– moral based on religious codes, especially on the authority of God or religious founders. Religion is not a good basis for universal ethics because many people today do not embrace any religious tradition or they may belong to religious traditions which are in conflict. To make matters worse we have the problem of balancing rights of expression for non-believers with religious sensitivities of believers ("the right-to-offend") [10]. The root of the problem is that most religious traditions contain universal claims about the validity of their own particular practices and world-views. Since religious traditions differ, so do their universal world-views which may lead to group conflicts.
- **Value-based ethics**- moral based on absolute human values, i.e. intuitively and emotionally recognizable aesthetical and moral norms (e.g. sanctity of life and property, freedom of expression, tolerance). This "ad hoc" hypothesis is designed for the purpose of justifying our existing moral views and though it seems appealing at first glance, it is often inapplicable because it is too general. For example, even though we may agree with the principle of sanctity of human life (as enshrined in UN Charter and similar documents) Life creates situations where this principle does not provide clear guidelines. What happens to 'sanctity of human life' in termination of pregnancies, euthanasia, death penalty, civilian casualties in war etc.?
- **Science-based ethics**- this type of ethics derives moral norms from the principle of biological evolution. The principle rationalizes ethical practices of species/social groups on the basis of their survival and reproductive success. The success ensures

> transmission of genetic information from one generation to the next. This principle seems straightforward, it is Science based, universally valid and "fair" (because natural selection operates on groups rather than on individuals and does so in a statistically unbiased way). The new discipline of 'sociobiology' or 'evolutionary psychology' expounds human nature and how it evolved as part of the evolution of biological species. Sociobiology was espoused by biologists ***E.O. Wilson*** and ***R. Dawkins***. Sociobiology proposed many interesting explanations for common aspects of human behaviour (competition, altruism, cooperation). Let us examine sociobiology in some detail because it is one of the very interesting derivatives of the Theory of Evolution.

Genes of an individual organism (genotype) determine reproductive fitness, health and other properties of this organism (phenotype) which competes with other organisms for resources in life and for reproduction opportunities. Similar genes occur in siblings of an individual and his descendants. Humans are social mammals so Nature & Evolution favour the behaviour of individuals who at their own personal risk improve the chances of survival of their siblings and descendants (and hence their common genotype). For example, this can explain altruism and sacrifice parents show towards their children. This is a subconscious, instinctive habit developed through evolutionary struggle. Altruism is selective, reciprocal and gradual i.e. it depends on the extent of relationship between individuals, on parental care and sacrifice, on group or clan cohesion. It is expressed as mutual help by members of the same group and as indifference or hostility to outsiders. Outsiders are often treated with indifference or hostility which leads to e.g. ethnic tensions within society, nepotism, xenophobia. Reciprocal altruism can also be explained by the same idea: mutual help is better for the survival of genetic pools of individuals than aggressiveness or indifference.

Altruism is a form of cooperation related to organisms with similar genotypes. However, cooperation can also exist between organisms which have different genotypes.

How did cooperative behaviour evolve?

Sources:
http://www.ogp.columbia.edu/images/photo_contest_2008/wenzhou_tandem_bike.gif
http://apps.detnews.com/apps/history/index.php?id=21

Fig. 36.

Figure 36 shows that cooperation is a matter of balancing gains and losses for individual members of the group. We can use tandem bicycle analogy. In the non-cooperating group some members gain (those who do not pedal) and some lose (those who do), but any individual gains in this group are much smaller than if all members were cooperating/pedalling. Cooperation thus arises through Evolution in a competitive world. Cooperation makes Evolution constructive, because it allows new levels of organization to appear which in turn enhances biological diversity [11]. The existence of competition alone (which is most people's perception of Evolution) would not generate much diversity in the biological world and would undercut the possibility for change. Evolution therefore rests on three principles: mutation, natural (i.e. environment driven) selection and cooperation [11].

Another interesting and common phenomenon which can be partly explained through ideas of sociobiology is the physical attraction between opposite sexes. For example, the sexual attractiveness of a woman is interpreted by Men as a visible and accessible (but not always reliable) indicator of her health and reproductive potential. Physical beauty implies healthy genes and thus a good chance of bearing healthy offspring. Human attraction to physical beauty of the opposite sex is thus also viewed as genetically driven; it is related to the probability of bearing healthy offspring [12].

source: http://www.adrants.com/mt335/mt-search.cgi?IncludeBlogs=1&search=ad%3Atech&x=0&y=0

Fig. 37.

With the spread of genetic testing and genetic databanks we may no longer have to rely on outward indicators (Figure 37); this may give more people a chance to meet suitable partners. In spite of the convincing arguments proposed by sociobiology regarding biological and evolutionary origin of human behavioural traits, the analysis of human beings as 'social animals' which exhibit no qualitative differences amongst themselves is inadequate [13] (as the quotation below humorously outlines). Sociobiology, in spite of protestations to the contrary, assumes the existence of an ontological entity which is genetically determined and has biological survival as its goal [13]. The importance of historical and cultural relations between humans is thus either neglected or even when acknowledged it is not considered in a systematic way and integrated with the analysis of biological foundations of humanity.

> *"Any scientist who has ever been in love knows that he may understand everything about sex hormones but the actual experience is something quite, quite different."*
>
> *K. Lonsdale*

Another problem for sociobiological basis of Ethics is that empirical facts cannot by themselves provide the basis for ethical norms. **It is impossible to deduce norms from mere facts!** Such attempts are an example of 'naturalistic fallacy' and can lead to oppressive and discriminatory social practices as we shall mention later in this Chapter. The fact that humans have originated (by the process of natural selection) from common ancestors as many animal species did, does not necessitate that we apply the same natural selection principle to human society. Sociobiology is an example of reductionist approach and of an extrapolation from Science to Ethics, so its claims need to be carefully scrutinized.

- **Community-based ethics** relies on current, 'politically correct' ideas about tolerance of many various traditions and customs, with each tradition purported to have its own ethically justifiable features ("multiculturalism"). This is not a good basis for ethics, although tolerance must be preferred to force. Certain moral norms and practices are unacceptable and cannot be tolerated under the guise of pluralism (e.g. racial cleansing practices in Nazi Germany, oppression during Khmer Rouge regime, religion inspired terrorism etc.). Furthermore, even the simple case of the "right-to-offend" pertaining to freedom of expression for non-believers vs. religious sensitivities demonstrates problems with the communitarian basis for ethics [10]. How can we apply 'politically correct' ethical norm when my freedom of expression is at the same time the source of offence to somebody else?

Patzig [9] suggested that a useful practical basis for ethics might be derived from the consideration of the effects our actions on others ("ethics of reciprocity").

"It is immoral to do something that with good reason, one would condemn in others if one were to be affected by this action."

Ironically, this prescription is similar to ethical recommendations proposed by several religions and philosophers long time ago. For example, see New Testament; Mathew 7:12. We are not born equal (see Chapter 1) since we evidently have different physical and mental abilities. This inequality makes establishing universal ethical norms difficult because the problem arises as how to best integrate a set of diverse human individuals into the functioning group. There are clearly various possible ways in which this integration can be done; each way represents a distinct ethical tradition expressed through a set of norms. The adequacy of particular ethical tradition cannot be easily assessed because the universally acceptable external reference against which such assessment can be made is difficult to define. Hence the 'ethics of reciprocity' seems to be the best recommendation at present. In short, the derivation of **absolute, universal** ethical norms remains problematic, because Ethics arises out of immensely complicated social interactions and the diversity of individual human beings.

Ethics in Science

Ethical code for scientific practice is a mixture of logical, historical and social ideas about how Science should operate. The code is not 'fixed' and some ethical ideals may change with the changing social environment. For example, new additions to the ethical code of Science like patent and property rights or conflict of vested interests in business companies have now been introduced.

However, ethics of Science is based not only on social influences and norms, but also on the universally valid Laws of Nature and practical experience regarding the most efficient way to organize the study of Nature. That is why Science should not conform to societal and political norms or fashions e.g. relativism, postmodernism, 'political correctness" and so on. There have been ill-advised attempts at social interference in Science e.g. introduction of 'German Science' against 'Jewish Science' in Nazi Germany 1933-1945 (see quotation below). The outcome was the impoverishment of German Science and the shift of World's scientific activity from Germany to USA.

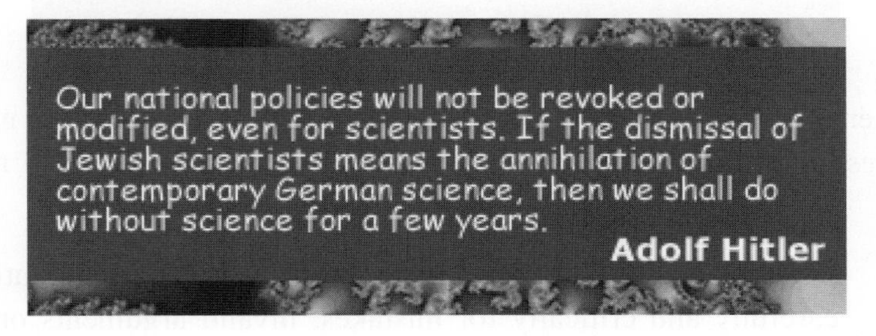

Our national policies will not be revoked or modified, even for scientists. If the dismissal of Jewish scientists means the annihilation of contemporary German science, then we shall do without science for a few years.
Adolf Hitler

Ethical standards pertaining to Science:

- **Originality and curiosity** → scientific studies should provide new and important information so that knowledge can advance; studies which are not new are not part of Science. Curiosity is the essential norm for Science even though it may lead to dangerous discoveries (e.g. in nuclear physics). Curiosity is not readily acceptable in other social activities/fields e.g. in business where trade secrets discourage curiosity.

- **Universalism** → there are no privileged sources of scientific knowledge. Scientific claims should be assessed purely on their own merits (evidence and logic), irrespective of the scientist's nationality, gender, race, religious and political persuasion or his standing in the scientific community. There is no absolute "authority" in Science as there is in religion or politics. Einstein has argued against the acceptance of quantum mechanics, because he considered it to be incomplete. Nonetheless, scientific community did not accept his arguments due to a large body of experimental evidence which supports the descriptions and predictions of quantum mechanics. Chekhov's quotation below clearly illustrates the universal nature of Science. Science was 'global' long before the economists and business people started floating the idea.

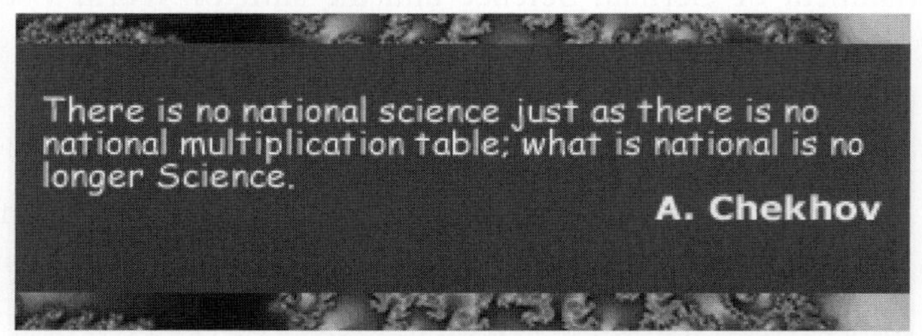

There is no national science just as there is no national multiplication table; what is national is no longer Science.
A. Chekhov

- **Scepticism** → all scientific claims should be scrutinized carefully and critically for mistakes, invalid arguments or even fraud. This scepticism is embedded in the "peer review system" (see Chapter 2) and ensures continuous scientific progress by eliminating wrong avenues of research and incorrect results. Once again, this is an ethical norm which is emphasized in Science much more than in other areas of human activity. Scientific investigations are always based on earlier results, and if these results are wrong, a lot of time and effort would be wasted in following false leads.

- **Communalism** → scientific knowledge is the property of everyone; it is not scientist's private property and should be transmitted to wider scientific community and to general public. Otherwise the cumulative progress of Science is impeded. Science critically relies on free exchange and discussion of ideas and results. However, some scientific research is performed under conditions of military or commercial secrecy by various organizations including those supported by Governments (see logos below).

sources:http://www.hypax.com.sg/images/client_dso.gif /
http://www.nictia.org.au/images/pagemaster/DSTO_black_vertical_logo_Custom.gif

- **Independence and freedom** → scientific knowledge should be free from political interference and scientists work best in the environment of social tolerance. The scientific truths are established through the consensus within the scientific community. This consensus depends on empirical evidence and logic and not on social circumstances or forces. There are no large lobbies in Science as in politics which promote their own special interests nor is the conduct of Science influenced by external political or economic influences. Please note that on the other hand, scientific research aims and directions may be influenced by external factors as was discussed earlier.
- **Open-mindedness and detachment** → scientists should work for the advancement of knowledge regardless of their personal

opinions or personal gain and be prepared to give up their beliefs in face of contradicting evidence. Nonetheless, the 'conflict of interests' may arise in scientific practice and violate this ethical principle (e.g. a scientist doing research on side-effects of a drug for the company which produced the drug and which employs him). Another factor which may sometimes work against this ethical principle is the psychology of scientists themselves (see quotation below). The ethical principle of open-mindedness and detachment must be abided by the scientists even though it may be a difficult task. This ethical principle is adhered to even less in other fields of human activity e.g. in business or politics.

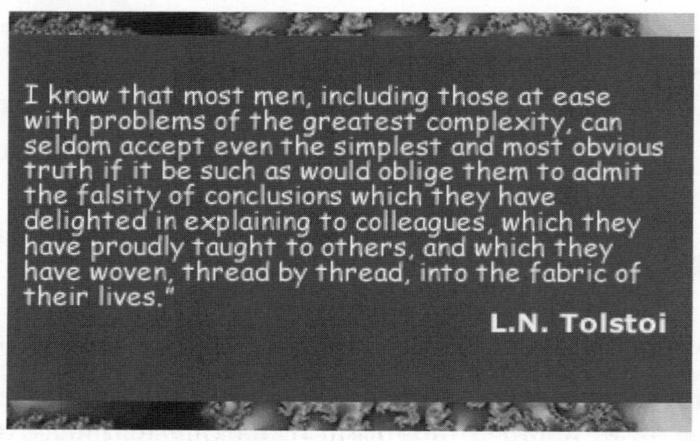

I know that most men, including those at ease with problems of the greatest complexity, can seldom accept even the simplest and most obvious truth if it be such as would oblige them to admit the falsity of conclusions which they have delighted in explaining to colleagues, which they have proudly taught to others, and which they have woven, thread by thread, into the fabric of their lives."
L.N. Tolstoi

- **Honesty and Integrity** → scientists, on average, are neither more nor less honest than other professionals or an average citizen. However, unlike in e.g. diplomacy, law or business, where one may assume that the other party is not necessarily honest (i.e. that the other party is lead by personal gain) this assumption is not acceptable in Science. Science has cumulative nature which means that scientists must rely on the results of their scientific predecessors and trust that the predecessors performed and reported their investigations honestly (that they did not forge their results). If this were not so, scientists would need to start all investigations from scratch which would bring Science to a halt. There are few legal penalties for fraud in

Science, but the offence against this ethical principle is taken very seriously and usually results in career termination for the offending scientist. Nonetheless, the misconduct (especially fraud) still does occasionally occur in Science, because some scientists are tempted by potential rewards. It is ironic that due to the peculiar character of scientific activity fraud in Science is usually quickly uncovered, unlike in other human activities. The reason for this is that the reported results are often checked by other scientists who work in the same field. The more important the reported results are, the more stringent the checking and attempts at reproducing the published results shall be. The example of 'cold fusion' (Chapter 2) is a case in point, even though Pons and Fleischman **did not fabricate** their results. Their results were so unusual and potentially useful (as a source of clean energy) that immediate large scale attempts were made to replicate and extend their work.

Misconduct in Science:

Misconduct includes fraud, plagiarism, and questionable research practices.

- Fraud can appear in the form of making up nonexistent results ("fabrication") or changing the existing results to support one's expectation ("falsification").
- Plagiarism is the use of someone else's ideas or results without due acknowledgment.
- Questionable research practices come in many forms, they are very common and difficult to detect; some examples are given below:
    - Failure to keep original research data or not keeping adequate research records.

- Conferring or requesting authorship on research articles on the basis of specialized services/contributions unrelated to the research being published. 'Name on the paper' syndrome!
- Refusing access to the peers of unpublished research data which support the results of published papers.
- Inadequate supervision or exploitation of research subordinates.
- Misrepresenting speculations as facts or releasing preliminary research via public media without allowing peers to check and reproduce the results. For example, in the 'cold fusion' case the scientists released their preliminary results to the media before publishing them in scientific research journals where they would have been subject to peer review scrutiny.
  - Selective research reporting (e.g. changing the hypothesis after data have been analysed -'post hoc hypothesis', reporting only data which support expected conclusions etc.)
  - Interference with fellow scientist's research work (harming other scientist's research for personal reasons)
  - Self-plagiarism (publishing the same results in two different journals, 'slicing' research results into several papers to boost publications number)
  - Conflict of interest (e.g. between advancement of knowledge and profit). For example, a scientist studying the environmental effects of pesticides funded by the company which manufactures the pesticides alters his research findings so as not to damage business interests by not reporting how harmful the pesticides really are.
  - Unethical research practices e.g. failure to obtain informed consent from human subjects on whom research is being conducted, using results of unethically conducted research.

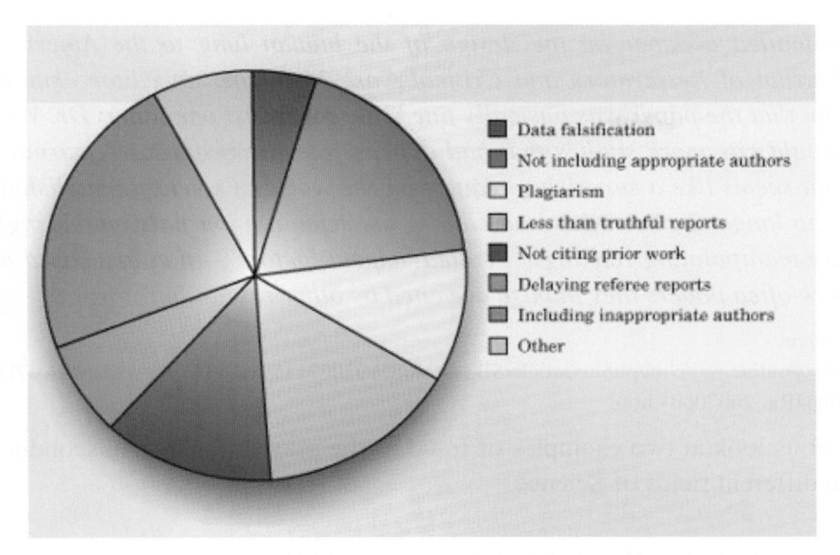

Reprinted with permission from: K.Kirby, F.A. Houle, "Ethics and the Welfare of the Physics Profession", Physics Today, vol.57, issue 11. Copyright 2004, American Institute of Physics."

The pie chart above describes the distribution of categories of ethical violations reported by junior members of the American Physical Society (APS). 39% of the junior members responded to APS ethics survey, citing one or more of these transgressions.

Here is another example of scientific misconduct but this time related to scientific publishing and not to individual researchers.

The excerpt from the article below in the Wall Street Journal, published on June 5, 2006, describes the unethical practice of some science journals. The journal editors encourage authors to increase the number of citations from that particular journal in their (author's) article to boost journal's impact factor.

'Dr. West, the Distinguished Professor of Medicine and Physiology at the University of California, San Diego, School of Medicine, is one of the world's leading authorities on respiratory physiology and was a member of Sir Edmund Hillary's 1960 expedition to the Himalayas. After he

submitted a paper on the design of the human lung to the American Journal of Respiratory and Critical Care Medicine, an editor emailed him that the paper was basically fine. There was just one thing: Dr. West should cite more studies that had appeared in the respiratory journal.If that seems like a surprising request, in the world of scientific publishing it no longer is. Scientists and editors say scientific journals increasingly are manipulating rankings -- called 'impact factors' -- that are based on how often papers they publish are cited by other researchers.'

source:
http://online.wsj.com/public/article/SB114946859930671119eB_FW_Satwxeah21loJ7Dmcp4Rk_20070604.html

Let us look at two examples of fraud as the gravest form of misconduct, in different fields of Science.

Example 1: Jan Hendrik Schön (physics) [14]

In 2002 Bell Labs in USA dismissed young nanotechnologist for falsifying data.

He has claimed to have produced various pieces of electronic circuitry from organic crystalline materials e.g. pentacene. His results were very exciting and promising which lead to his work being published in top journals: Science and Nature; he was even touted as a candidate for the Nobel Prize. His work involved several senior scientists as co-authors on his publications. Schön's case is a good example of a failure of peer review system. His work not only passed the review scrutiny in top journals, but also exemplifies the marketing/business considerations by editors of scientific journals who claimed that Schön's work was so novel and exciting that it had to be published. Furthermore, the guilt of scientific misconduct applies not just to Schön himself, but also to his senior collaborators who accepted co-authorship without thorough analysis of Schön's data. What were the driving forces behind these breaches of ethical conduct? Perhaps there are several related factors which can be conflated under the same name: fascination with quick success at any cost! This human desire has driven several people in this story (Schön, the journal editors, referees, Schön's colleagues) to go beyond the ethical boundaries of Science. It should be emphasized that it is not only Schön who should shoulder the blame, but other members of the scientific community as well.

Example 2: Hwang Woo-suk (molecular biology) [15] *South Korea's Hwang Woo-suk was celebrated as a national hero when, in 2004, his research team claimed to have successfully cloned a human embryo and produced stem cells from it, a technique that could one day provide cures for a range of diseases.*

But allegations that he used unacceptable practices to acquire eggs from human donors (members of his own lab), then faked two crucial pieces of research into cloning human stem cells, have left his reputation in tatters. He was dismissed from his academic position and taken to court for fraud. This case is interesting because it highlighted the national involvement in Science (Woo-suk was a national hero), legal implications (he was prosecuted) and the unethical use of collaborators in his research (egg donations).

Why does misconduct in Science matter?

Science progresses because scientists produce honest research in an honest way. Honesty ensures public trust in Science (which is important for social support of Science) and attracts able young people into scientific careers. Dishonesty damages all three goals. The scientific community punishes misconduct by withholding research grants or by terminating the culprit's scientific career.

However, there is no uniform way in which scientific misconduct can be dealt with; there is no 'science court'. In some cases, misconduct can be hushed up so as not to damage reputations of institutions or individuals. 'Whistle-blowing' exposes the perpetrator to professional risks (career difficulties, job loss etc.).

Reflection:

Are the ethical values of Science and Society identical? If not, why not? Can they possibly be identical?

Hint: consider what are the reference points for scientific and for social ethics

Societal support for modern Science

We have already mentioned that Science is a social activity which requires social support. This is especially true of modern Science which requires expensive and sophisticated experimental equipment. However, the more Science advances less possible it is for general public to understand it. We thus have a paradox: the public supports research which it does not understand. The quotation below highlights some of the problems stemming from this paradox.

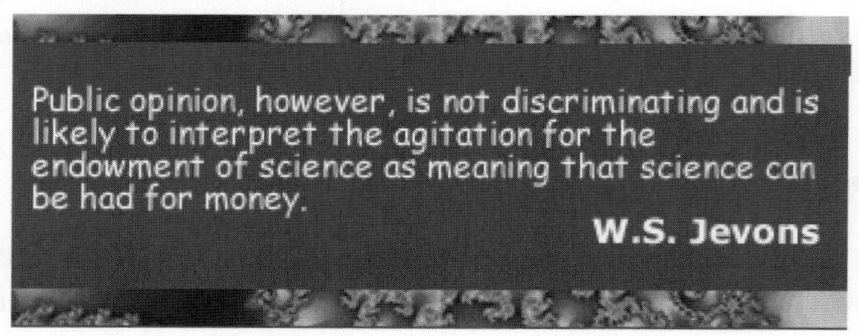

Public opinion, however, is not discriminating and is likely to interpret the agitation for the endowment of science as meaning that science can be had for money.
W.S. Jevons

The societal support for Science is today considered to be in the national interest, but it was not always so, as the concept map below indicates. Amateur scientists had freedom to choose their research objectives and were not restricted by external funding. This was of course only possible when requirements of scientific research were modest in terms of equipment and cost.

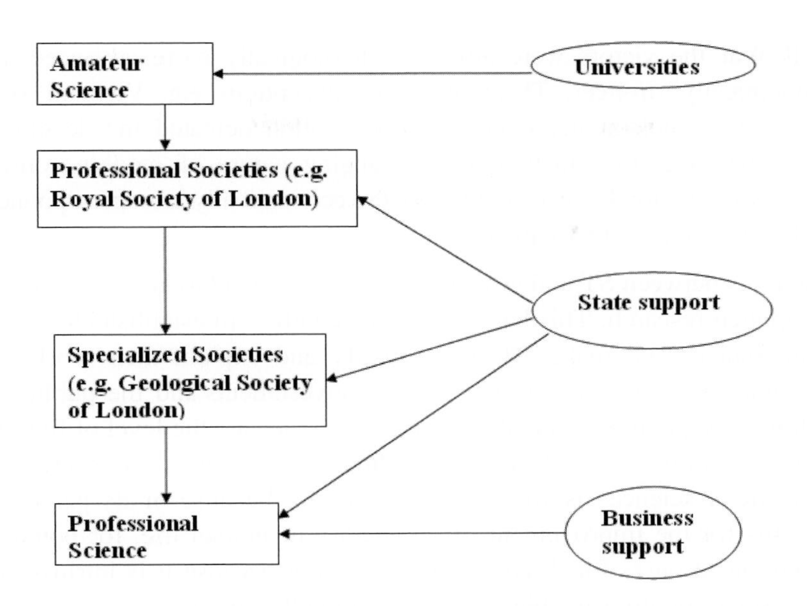

Initially, all scientific activities were financed from private resources of scientists themselves or by wealthy benefactors (kings, nobles etc.). F. Bacon was well ahead of his time when he suggested that State should support systematic scientific research and that its results should be systematically exploited for practical, social benefits.

In 17th century, the European states funded the establishment of scientific societies (e.g. Royal Society in UK) and such societies provided scientists with meager support. All the experimental equipment was then 'home made' and often crude; many scientists were performing both theoretical and experimental studies. Newton was not only a great theoretician (founder of physical theories and of calculus), but also a skilled equipment builder (optical instruments). From 19th century onwards the main financial support for Science came from state funding and business capital. World Wars were total wars and convinced the public and governments that Science and Technology (ST) is crucial for survival and victory. Fate of modern wars depends to a great extent on the economic and ST capabilities of the country involved. In modern wars like WWI, WWII and

Gulf War the victors were always technologically more advanced and economically stronger. There are some exceptions e.g. Vietnam War, where a technologically underdeveloped nation defeated the developed one. The exceptions may indicate changing nature of modern conflict away from the total war which favors the economically stronger opponent with more advanced ST capability.

The links between ST and government financing lead to greater emphasis on applied research. This has compelled scientists to establish closer ties with Society and make efforts to justify and explain their work and potential benefits to the Society. Both Governments and the public are willing to support Science; the contentious issues are the level of funding and the allocation of research directions/priorities. Research in biomedical sciences is well funded primarily because of its perceived benefits for the improvement of the quality of human life, for potential invention of cures for diseases and aging, not because it is intrinsically more scientifically valuable than research in other fields.

Modern scientists do not live in an ivory tower (aloof from real concerns), but have their work publicly scrutinized and they have to compete with other scientists and other social programs for limited funding resources. For example, scientists had to acquire the new skill of 'grant application writing' and convince funding agencies of the merit of their scientific proposals. In fact a lot of scientist's time is spent writing such proposals. This necessarily reduces the time and effort dedicated to supervising research collaborators and to developing new research ideas. Why do scientists then do Science?

Motivation of Scientists

Scientists generally have two types of motives for doing Science:

- **Personal motives** -curiosity, pleasure in doing research, desire for scientific reputation and influence in the scientific profession or outside of it.

- **External motives** -public fame, desire to find practical applications of scientific knowledge, need to get funding for research or to make profit from it, desire to influence public policy, or to gain social prestige.

sources:
http://www.wpclipart.com/cartoon/people/women_cartoons/another_tricky_day_for_Mom.png.html

Fig. 38.

There is a *gender* problem in Science because women are underrepresented. The principal reason for underrepresentation is the traditional social role of women and related prejudices about women's mental capabilities for natural sciences (Figure 38).

Most scientists do Science for a combination of motives, but usually not for the benefit of a higher pay. Most scientists are not 'nerds' even though they may not be good communicators. Sometimes the desire for a reward or fear of failure may lead them into fraudulent practices, but the penalty (within the scientific community) is very high, because Science can only operate if scientists are honest in their work. Scientific community has its hierarchy, but while this hierarchy does not influence the final outcomes of scientific investigations, it may have influence on the success or failure of individual scientists' careers. It is interesting to note that the level of satisfaction of scientists with their careers depends not only on their own individual efforts, but also on the type of institution where they work [16].

Scientists in Society

Scientists interact with Society in various ways:

- By devising new technology which can be in widespread use
- By educating general public through popularisation of Science via media interviews, special science programs, popular science books etc.
- By participating in legal processes (e.g. use of DNA evidence in criminal trials, as expert witnesses)
- By espousing specific ethical and cultural values in public and as private citizens (values of critical inquiry, rational argument, integrity etc.)
- By influencing public policy directions and decisions when acting as government advisors or industrial consultants.

In spite of the contributions of scientists to Society, many stereotypes and prejudices about scientists as human characters remain. Some are listed below.

Some stereotypes of scientists are uniformly negative or uniformly positive while others fall in-between these two extremes.

- **Evil alchemist** (e.g. Doctor Faustus, see below) - he is engaged in secretive, illegal research, proud, rebellious, arrogant, power-crazed, insane in trying to transcend human limitations. This stereotype is a parody of scientists because nowadays scientists cannot afford to be secretive even if some of them do suffer from the other traits mentioned. The same traits can however also be found in personalities prominent in business, media or politics.

The image of Faust still has the heroic, romantic, promethean charm [17] as it depicts the straining of human intellect against the limitations of human knowledge and Life. Few scientists today belong to this stereotype in our distinctly unromantic age. This stereotype pictures a tragic romantic hero who may be bad, but who is also pitiful in his passionate and failed attempt to comprehend the absolute Truth as the secret of Life.

source: http://www.amshakes.org/shows/faust/artwork.jpg

- **Noble scientist** (e.g. Newton) – he is an altruistic idealist, engaged in the pursuit of common good, seeking to explain rather than to mystify. He is humble and delights in bringing harmony into chaotic Nature. He is the symbol of triumph of human reason in the search for understanding of Nature. The historical Newton did not conform to this ideal [18] which again demonstrates that this stereotype has more to do with human aspirations and perceptions than with reality. This stereotype presents scientists in unreservedly and undeservedly positive light.

source: http://www.wpclipart.com/famous/science/Newton/Isaac_Newton.png.html

- **Obsessive/absent minded professor** (e.g. Einstein) - Individuals who are too engrossed in their research. They wear non-matching pieces of clothing, forget their own weddings or other social engagements, do not take family responsibilities seriously and are driven by curiosity from one crazy project to another. This nerdy stereotype fits Hollywood perceptions more than it does reality. The people with such personalities can also be found in other professions (e.g. artists). This is essentially a comic portrait of scientists which puts emphasis on a funny side to increase its market appeal.

source: http://www.wpclipart.com/famous/science/Einstein/Einstein_2.png.html

- **Inhuman rationalist** (e.g. J.R. Oppenheimer) – he is pursuing purely rational, intellectual study of Nature regardless of social harm and abandons human relationships. He/she disregards the cost and consequences of his/her inventions. This is again a stereotype because although Oppenheimer's work did lead to nuclear weapons he became a great proponent of nuclear non-proliferation and World peace afterwards.

source:
http://www.sfgate.com/blogs/images/sfgate/goldberg/2009/04/12/robert_oppenheimer_3.jpg

- **Adventurer hero** (e.g. Star Trek character) – this is an attractive, but simplistic, idealized, media designed and driven portrait of scientists. Scientists according to this stereotype are adventurous, courageous, morally superior people. This description belongs to the realm of science fiction. It is much more about heroes (who were portrayed throughout history as brave, self-sacrificing, reckless, patriotic) than about scientists. In this stereotype a typical hero is only thinly disguised as a scientist to increase the market appeal of the narrative.

source: http://www.techfodder.com/2009/07/07/StarTrek.jpg

- **Out-of-control** (e.g. Frankenstein) - Individual who refuses to foresee and accept responsibility for disastrous consequences of his research and as a result brings calamity on himself and others. This stereotype of scientists is essentially negative and similar to the "evil alchemist", but without the former's noble passion for deeper, ultimate Truth. This stereotype shows no sympathy for its subject i.e. scientists deserve what they get when they become 'hoisted on their own petard'.

source: http://www.nde.state.ne.us/ss/irish/franken.gif

Scientists, like any other social group comprise people of various personalities and interests. Many scientists are engaged in writing (e.g. authors of the play 'Oxygen' by Djerassi and Hoffmann) or other non-scientific activities. While each stereotype contains some similarity to the living scientists none is to be taken literally. These stereotypes are, like all stereotypes, actually caricatures.

Science, Technology and Applied Science

These three activities are often considered to be almost identical due to their inter-connectedness in modern Society. They do however have distinct identities.

Science seeks to understand the World within the most comprehensive and general framework.

Technology seeks to control our immediate physical environment. Technology comprises a set of tools, 'know-how' designed to make things happen according to our wishes. Technology is a very important bridge between Science and Society.

Applied Science has several meanings:

- Use of scientific results to achieve practical ends
- Scientific study of a process or phenomenon of practical importance
- Development of novel engineering techniques based on current Science

Science(S) and Technology (T) have developed independently, but not in isolation e.g. steam engine (T) was developed without scientific insight, but it lead to the development of scientific discipline of thermodynamics (S). Many public concerns about Science in fact have more to do with Technology than Science.

Differences between Science and Technology include the following:

- S is universal; T is particular
- S and T have different criteria; in S what can be done is often done right away; in T what can be done may not be done, depending on the perceived economic benefit (e.g. solar power technology while possible is not used yet, human cloning is constrained by ethical concerns)
- S has continuity over time; T does not
- S comprises intangible knowledge; T comprises devices (tangible objects)
- S is open, publicly accessible and cannot be easily controlled; T is secret and may be controlled (e.g. in military or industrial research)
- Scientists have strong links within their scientific community (expressed through peer reviews, publications); Technologists do not often have such strong links and are more influenced in practice by their clients and employers.
- All scientific research cannot be directed towards solving practical problems (as political and business leaders demand), because the applicability of basic research is unpredictable. Technological research is by definition aimed at practical use.

These differences between S and T are expressed nicely in the quotation below.

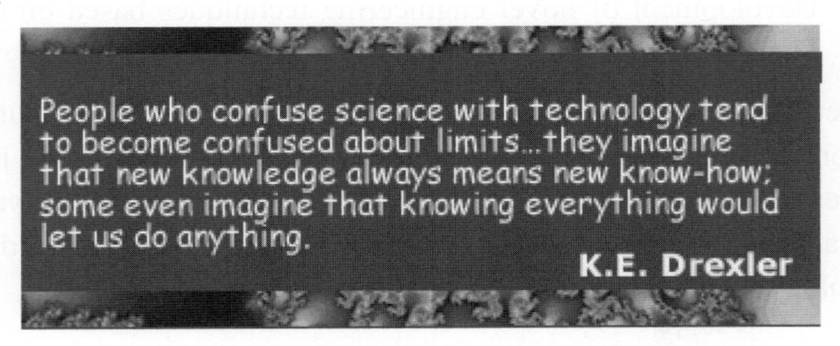

People who confuse science with technology tend to become confused about limits...they imagine that new knowledge always means new know-how; some even imagine that knowing everything would let us do anything.
K.E. Drexler

Science impacts on Society

Science and Society interact (directly or via new technologies) and this interaction is the source of many interesting ethical and social changes and challenges. The exchanges between Science and Society are reciprocal. Science stimulates Technology which affects Society which in turn influences the directions of both scientific and technological advances by setting their goals and priorities. The discussion of consequences of any particular scientific discovery in this book is **NOT meant to imply any ethical judgements!** The sole aim is to describe the chain of causally connected events.

Science can influence social values and ethics and transform Society either via new technologies or by generating new ideas which originate from new scientific discoveries.

ST can give us new abilities and understanding, but it cannot make value judgments on our behalf. ST raises new questions, dilemmas and profound challenges. Some examples are mentioned below.

- **New technologies lead to new abilities:** Nuclear weapons and other weapons of mass destruction (WMD) are the best known, but not the most important example of Science-Society interactions. The striking images below have become part of our 'cultural and historical landscape' (Figure 39).

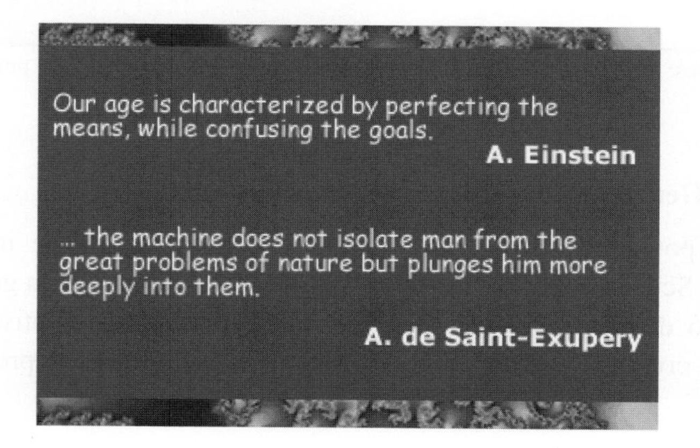

Our age is characterized by perfecting the means, while confusing the goals.
A. Einstein
... the machine does not isolate man from the great problems of nature but plunges him more deeply into them.
A. de Saint-Exupery

sources: http://www.wpclipart.com/weapons/nuclear/Nagasaki_bomb.png.html

Fig. 39.

- **Genetic engineering** (disease cure, cloning)

This is potentially much more important example of the interaction between Science and Society, because unlike weapons (which govern our ability to destroy) this scientific discovery affects the creative side of human civilization (family, friendships, emotions, reproduction).

Destruction is in itself a dead end while creation is open-ended, can follow many possible paths and lead to many possible outcomes.

- **Contraception and artificial fertilization** lead to profound changes in social habits, to the separation of social and reproductive roles of sex, to smaller families, to increase in the frequency of casual sex, to drop in birth-rates and subsequent aging of population in economically developed countries. None of these consequences were foreseen or intended and should not be attributed solely to the discovery of contraception nor do these comments represent an ethical comment on the contraceptive 'pill'. (Figure 40)

Other examples where Science becomes strongly intertwined with social values and considerations include:

- The discovery of tests for genetic diseases which raises many moral dilemmas: whether the patient should be informed, whether the doctor has the right to withhold that information, should the life insurance companies have access to that information, should the pregnancy be terminated if genetic defects have been discovered etc. Science provides the challenge by providing new information, how to use it depends on social and psychological considerations.
- Medicine uses many technologies for Life prolongation and support. Some arising concerns are: at what point should the terminally ill be allowed to die by withholding life support, how much resources should be devoted to such support vs. other social priorities like improving care for young children, should the biomedical research be directed towards finding cure for diseases of the elderly minority in the developed countries (Parkinson's, Alzheimer's etc.) or should it be focused towards finding cures for widespread diseases of the young majority of the World population in the developing world (malaria, tuberculosis). The decisions regarding these issues are based on political and economic criteria

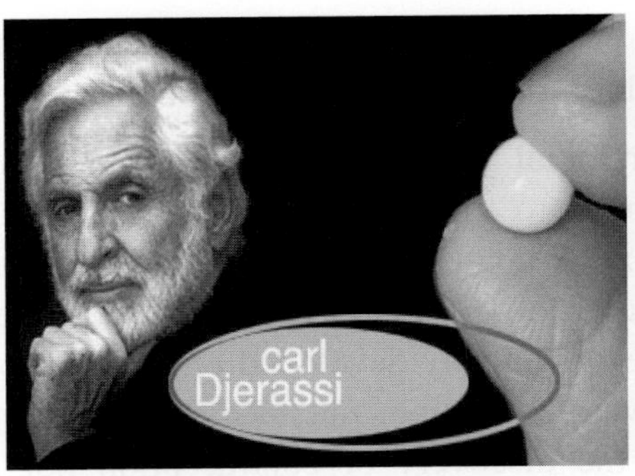

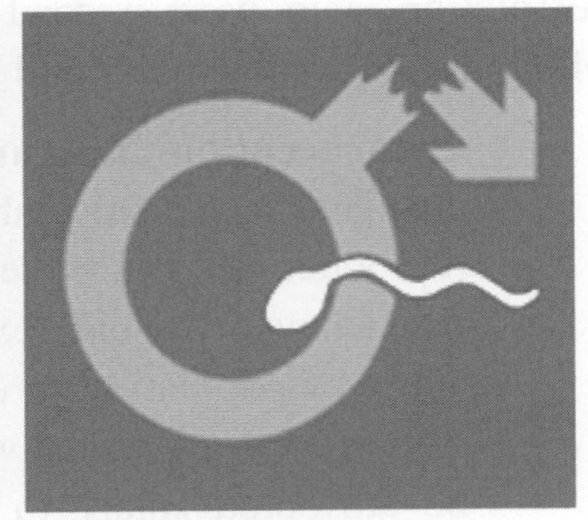

sources: http://intransigeants.files.wordpress.com/2009/01/pilule-carl-djerassi.jp
http://www.broadwaytovegas.com/taboos.jpg

Fig. 40.

(developed countries can afford high cost of medical research) rather than on scientific grounds. Science can of course pursue either direction of research.

- Chemistry has developed many new fertilizers and pesticides in order to feed the rising world population. However, these benefits are tied to costs in the form of environmental pollution expresses through toxic effects of pesticides and herbicides on human and animal life.
- The developments in nuclear physics lead to new weapons. Is mass killing of civilians justified? Are the resources spent on weapons research wasted? These ethical issues and dilemmas continue to this day (e.g. Iran's nuclear programme).
- Development of new technologies which drives globalization, often leads to the restructuring of economy with considerable social dislocations (job losses, work overload), marginalisation of some groups in Society (urban poor) and increased disparity in the rates of economic development in the World.

New self-image of Mankind [19]

sources:
http://www.wpclipart.com/famous/science/science_3/Nicholas_Copernicus.png.html

The *'1st scientific revolution'* (Copernicus and Newton) demonstrated that Universe is homogenous and all of its parts subject to the same laws. This revolution did not however go as far as suggesting that even Life as the most complex known phenomenon should be explained by the same laws as those valid for inanimate matter. The dualist view of the World postulating mind/spirit and matter as different entities therefore persisted. Dualist view held that complex phenomena like human life must have complex, non-material causes (e.g. 'intelligent designer', God etc.). This view also tacitly assumed that Life, especially human life, has a distinct purpose.

The replacement of geocentric with heliocentric model of Solar system by Copernicus had not only scientific impact, but also wider social consequences because it hinted that Mankind does not have the central role in the physical Universe as many religions have claimed. This scientific advance naturally created a sense of insecurity and doubt as the quotation below suggests (from ref. [20], pp.360)

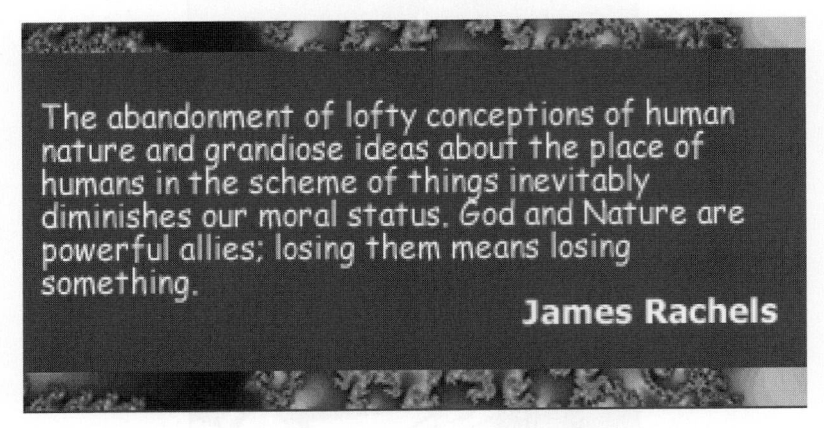

The abandonment of lofty conceptions of human nature and grandiose ideas about the place of humans in the scheme of things inevitably diminishes our moral status. God and Nature are powerful allies; losing them means losing something.
James Rachels

The '2nd *scientific revolution*', initiated by Darwin (see photo below) and his followers extended the validity of natural laws (set forth by the 1st *revolution*) by suggesting that even complex phenomena related to living organisms can be explained by physical laws operating in a purely material Universe. This monist, materialist theory has dispensed with purpose as a driving influence in Nature. Darwin's theory showed that teleological explanations are ***not necessary*** for understanding the origins and the existence of Life. The diversity of species can be created through natural selection. Natural selection is expressed through environmental constraints ('boundary conditions' in the language of dynamical systems) imposed on a given population and can take place without the intervention of intelligent agents. Natural selection is thus akin to 'invisible hand' of the market which according to Adam Smith, guides economic activity without outside rational interference.

The Theory of Evolution (TOE) had even more profound consequences for human self-esteem and place in the Universe than the 1st *revolution*. It had profound social impact because it demonstrated the continuity of all living species; it removed the perceived qualitative barrier between human beings and the rest of animal species. Darwin's book, *"On the origin of species"*, published in 1859, proposed that all biological species evolved through processes which included variation, natural selection (expressed through competition which leads to fitness) and adaption to the environment. The evolutionary process operates only on population of a particular species, not on individuals. This idea has also challenged the perception of importance and uniqueness of individual human beings. Nature truly does not have favorites.

TOE is an example of reductionism where a vast variety of phenomena have been 'reduced' to simple, general principles of mutation, natural selection and development. This is not surprising since scientific theories always have this task of compressing the available information by establishing appropriate general principles and patterns.

source: http://www.wpclipart.com/famous/science/Charles_Darwin_2.png.html

Darwin in his seminal work did not make any suggestions about ethical and social implications of his theory. This was in line with the practice of Science which proceeds carefully when making extrapolations beyond the range of validity of available empirical data. Nonetheless, Darwin's silence was an 'open invitation' to careless thinkers to try and deduce such implications from his work, often using unscientific reasoning or oversimplifications. Darwin's ideas or rather some of the implications deduced from Darwin's work had great social impact. They have challenged previously established views of the World and created controversies and debates.

The opponents of TOE have tried to show that TOE is not true in the absolute sense, that TOE is just a theory. We have discussed this position in Chapter 2 when discussing pseudo-science. Science does not claim to proffer absolute Truth about the World (because we can never gather all possible evidence), but this does not imply that no knowledge is possible. If, as TOE opponents suggest, knowledge can never be certain this would then apply in equal measure to their own negative views about TOE!

The controversies about TOE have little to do with the Theory itself or its empirical underpinnings which are well established. The open questions regarding TOE are related to extrapolations that were and are being made from TOE. Examples of such questions are: is dualistic view of the World (spirit-matter) plausible, does the understanding of our evolutionary origins shed light on human nature, do evolutionary origins govern and constrain human behavior and if so, should TOE be used in ethical decisions. These are very important questions and Richards [19] has performed careful logical analysis to examine whether inferences often made from TOE are *logically* justified. Surprisingly, she found that they are NOT. The only exception is the fundamental difference between materialist and non-materialist World views which seem irreconcilable. She has shown that many controversies involving TOE are based on perceptions which we have about ourselves and which cannot be affected by the progress of Science. In other cases, the controversies are fuelled by problems which are not scientific, but related to philosophical interpretations or logical misunderstandings.

Still, the fundamental difference mentioned above ('irreducible gradient' to borrow the expression from Schneider and Sagan [8]) drives and shall continue to drive 'Darwin wars'. This difference of viewpoints appears in many guises e.g. as reductionism vs. holism, monism vs. dualism and stretches in time back to antiquity. In the final analysis, this clash of viewpoints is a consequence of "algorithmic compression of information" which we perform when doing Science. The compression can be achieved in different ways and leads to different world-views. The compression is of course necessary if we are trying to describe a large complex system of interacting entities, especially if we ourselves are one of the entities in the very same system. However, the compression also necessitates making approximations about the system under study.

The examples of ethical problems and confusions which have been associated with Darwinism include:

- *TOE concepts have been linked to social issues and were assumed to justify right-wing political and economic systems*

There are many examples from politics and business where we can trace the influence of Darwin's ideas.

Political ideologues with hidden social agendas used TOE to support practices of 'social Darwinism' (which justifies social and class inequalities by the evolutionary struggle and survival) or 'eugenics' programs (selective human reproduction in order to breed 'superior' human race) which advocated sterilization or killing of deformed, disabled, mentally retarded members of society or of 'racially inferior' nations (genocide in Nazi Germany). It is no coincidence that the book by Hitler, one of the best known ideologues of racial 'purity' and ethnic cleansing is called *Mein Kampf* ('My struggle').

Figure 41 shows propaganda poster (with translation) aimed at swinging public opinion against medical support for mentally disabled and the cover of Hitler's book.

(Translation of the left poster: "60,000 RM is the cost of this man with an inherited disease for the duration of his life to the nation community. Comrade, this is also your money! Read the New Nation, the monthly journal of the office for racial policy of the NSDAP".)

sources:
http://media.photobucket.com/image/eugenics%20nazis/Primate_bucket/nazi_poster2.jpg
& http://truereligiondebate.files.wordpress.com/2008/04/meinkampf.jpg

Fig. 41.

The quotation below (ref.[20], pp.303.) highlights simplistic, pseudoscientific extrapolations and delusions of Nazi regime and proponents of similar ideas. The surprising part in the quotation is Hitler's comment about Christian religion as a rebellion against natural law. Christianity in his quotation appears to be a defender of human values. This is interesting because such depiction of religion is at odds with some streams of atheist thinking which portray religion as a dehumanizing force.

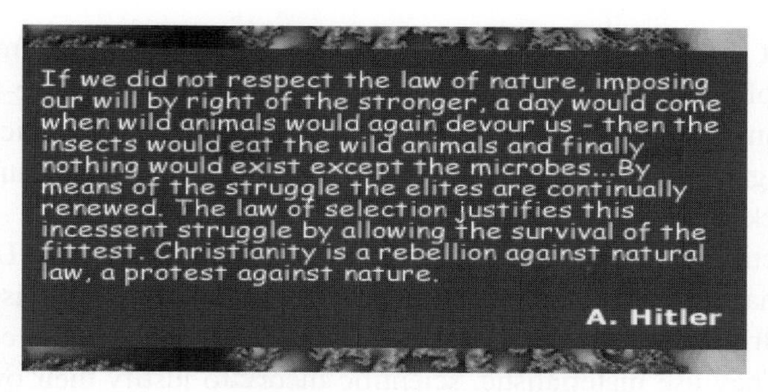

If we did not respect the law of nature, imposing our will by right of the stronger, a day would come when wild animals would again devour us - then the insects would eat the wild animals and finally nothing would exist except the microbes...By means of the struggle the elites are continually renewed. The law of selection justifies this incessent struggle by allowing the survival of the fittest. Christianity is a rebellion against natural law, a protest against nature.
A. Hitler

Another interesting quotation (below; from ref.[20], pp.177) exemplifies similar views in the field of economics.

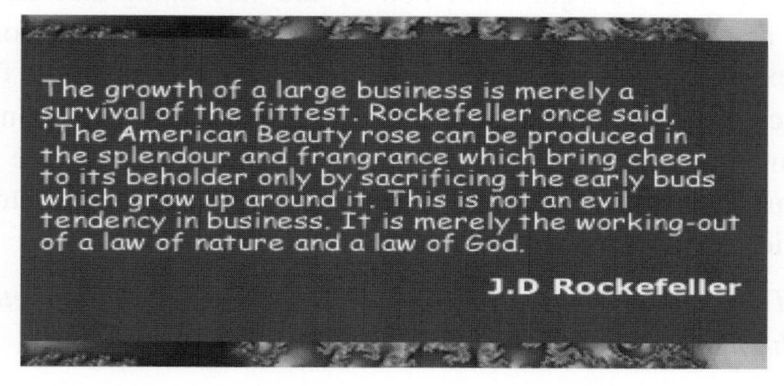

The growth of a large business is merely a survival of the fittest. Rockefeller once said, 'The American Beauty rose can be produced in the splendour and frangrance which bring cheer to its beholder only by sacrificing the early buds which grow up around it. This is not an evil tendency in business. It is merely the working-out of a law of nature and a law of God.
J.D Rockefeller

It is not surprising that politics and business may resort to Darwinian sounding concepts as justification of their policies since in both fields we can witness cut-throat struggles for supremacy. For example, in UK, the 1970s economic crisis was attributed by conservative politicians to the crisis of overblown welfare state. The National Health Service (NHS) was reformed, since it was considered to be too hospital-centered, curative-based and therefore 'too expensive' to run. Competition was introduced through private health care systems. The health care thus became considered as just another service delivered to the public, with patients treated as consumers. In such a system, the wealthiest have access to the best care, while others do not. This is still the problem in USA healthcare system where reform has been initiated by US president Obama. The fact that Darwin had nothing to say about such issues does not seem to perturb conservative thinkers. Furthermore, if we apply Darwin's evolution to Society literally, all traditional conservative values and human rights will be swept aside. There can be no human rights, individual freedom or sanctity of private property and life amongst the competing gene-machines [19]! Many conservative thinkers try to merge principles of ethics derived from religion with materialistic World view as exemplified beautifully by the Rockefeller's quotation above. Rockefeller did not see any contradiction between laws of Nature, expressed through Darwinian evolutionary struggle and the laws of God!? A similar example is given by former British PM Margaret Thatcher who said: 'Choice is the essence of ethics'. They use materialistic, scientific theory to justify their own social preferences and commit 'naturalistic fallacy' in the process. Ironically again, Hitler (see his quotation above) did not think that this merger between materialism and conservative, Christian ethics was possible. A similar example of incoherent reasoning which attempted to forcefully merge Science (considered intellectually respectable) with conservative ethical imperatives was published [21]. While one may agree with the ethical imperatives mentioned by Freeman [21], the connection to Science in the article is spurious and politically motivated.

- *TOE concepts were assumed to invalidate human ethics and human capacity for altruism*

TOE concepts raise awkward questions regarding human ethics. Since we are not born equal nor are we equally biologically fit, why should we treat every human person equally? Should biological fitness, in the World determined by evolutionary processes, not take priority in deciding how to treat individual members of human society? There are two reasons why such application of TOE concepts is misguided. Firstly, individual differences between humans are not entirely genetically determined [7]. We know this because genomes of many different biological species have similar sizes. For example, the genomes of humans and apes are very similar yet we are clearly very different from the apes.

Secondly, the objection regarding the collapse of Ethics as a result of Darwinism is based on the notion that without God there is no moral truth or moral standard. This opinion has been challenged [19]. Moral standards and ethics are necessary if human society is to function and produce advanced sophisticated structures e.g. civilization. This comment does not imply that Evolution process has deliberately designed humans so that we can build our civilization. However, once the Evolution produced organisms sufficiently complex to build advanced social organization, that organization (inclusive of ethical norms) must have arisen naturally from mutual interactions between advanced organisms under environmental constraints.

Finally, Evolution does not preclude altruism; it in fact supports a limited degree of altruism because without altruism higher, more organized forms of Life cannot evolve [11].

- *TOE concepts are assumed to propose deterministic gene-machine view of Nature, they preclude freedom of will and capacity for individual responsibility*

Some scientific thinkers did discuss free will by reference to physics which portrays humans and Society as machines ('clockwork Universe') where everything is predetermined and the free will does not exist. TOE was again used to justify certain views about human society. *Behaviourists* claim that free will is an illusion and that we are governed by past and present sensory stimuli. *Sociobiologists* (Wilson, Dawkins) claim that there are no qualitative differences between humans and social animals and that we all have similar patterns of behavior which are genetically determined (genetic determinism; new God?).

source: http://cognitionpress.wordpress.com/

In the sociobiologists' view (as derived from TOE), the primary goal of human action becomes biological survival of human species and is reflected in the preservation and transmission of genes. Sociobiology (a.k.a. evolutionary psychology) disregards the role of historical-cultural factors and relations within the human society. The quotations by Dawkins and Wilson below exemplify such arguments. Has genetics become the new Faith of modern age which can provide all kinds explanations for our behavior? We already have genes for obesity, anger, truancy......!

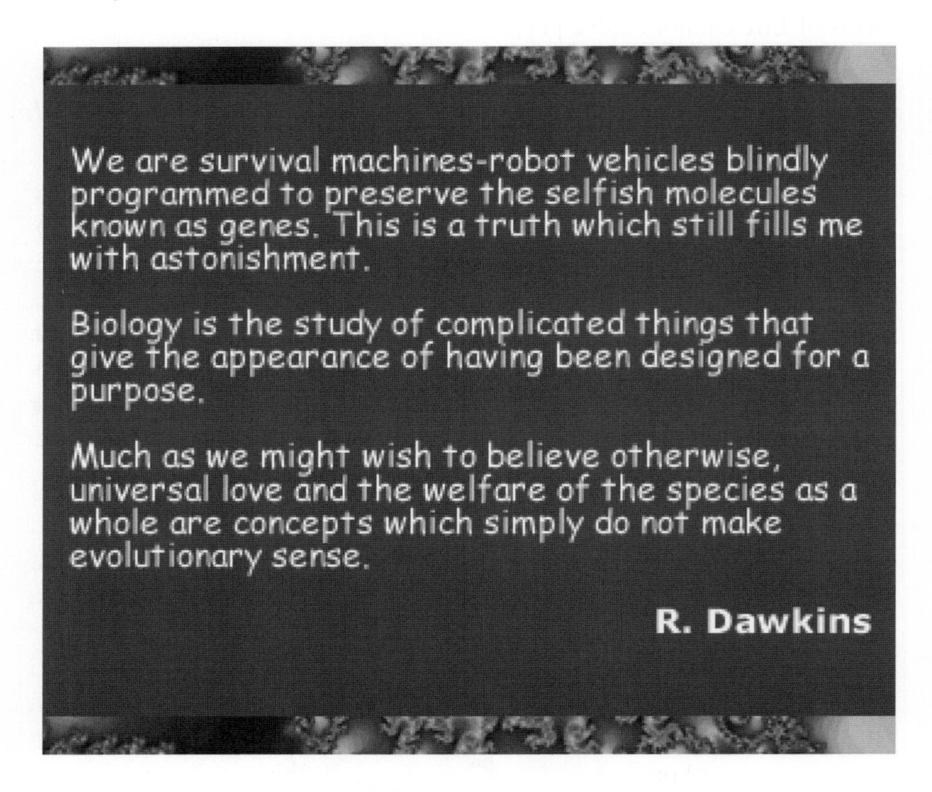

We are survival machines-robot vehicles blindly programmed to preserve the selfish molecules known as genes. This is a truth which still fills me with astonishment.
Biology is the study of complicated things that give the appearance of having been designed for a purpose.
Much as we might wish to believe otherwise, universal love and the welfare of the species as a whole are concepts which simply do not make evolutionary sense.
R. Dawkins

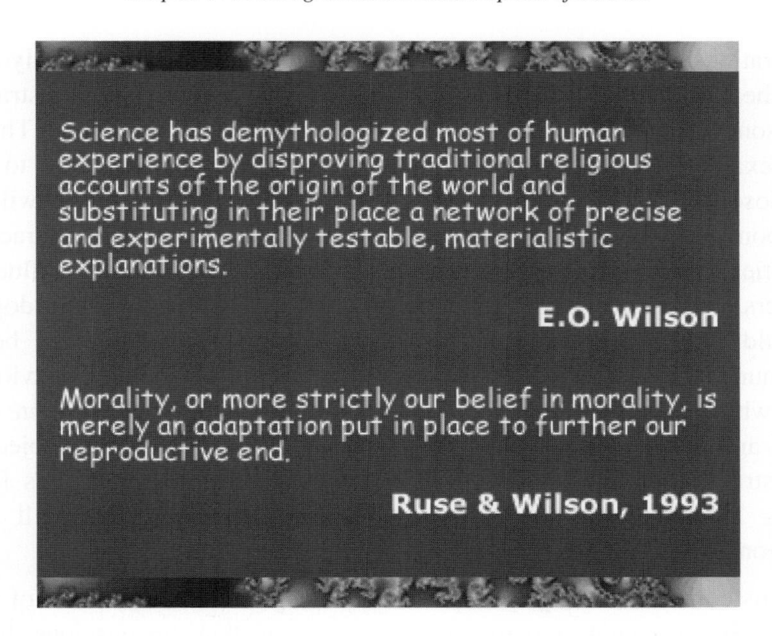

Science has demythologized most of human experience by disproving traditional religious accounts of the origin of the world and substituting in their place a network of precise and experimentally testable, materialistic explanations.
E.O. Wilson
Morality, or more strictly our belief in morality, is merely an adaptation put in place to further our reproductive end.
Ruse & Wilson, 1993

The statements by Dawkins and Wilson are extrapolations and exaggerations often used in public debates between atheists and religious believers [24]. Unfortunately, the exaggerated nature of such claims does not make them convincing or fruitful, it demonstrates lack of understanding of Religion and is scientifically unsound [24].

Human free will and personal freedom are ideals cherished by liberal democratic societies; they represent one of the pillars of their social and political thinking. An interesting illustration [22] of the tenacity of free will notion is the so called 'war on drugs'. The members of 'free societies' which are the largest consumers of illicit drugs, do not ponder a simple question: Why do so many members of free, advanced and affluent societies crave drugs (the demand side) in spite of being aware of their detrimental health effects? Market laws clearly state that there is no supply without the demand. Perhaps this question is too uncomfortable because it suggests that economically advanced, free societies have very fundamental problems in spite of their democratic character and embrace of free will. Instead, costly and ultimately futile operations are launched by developed countries to stifle the supply side in the developing countries. The belief in the sanctity of unrestricted personal freedom prevents rational analysis of the whole problem. This is an example of the vagueness and confusion of issues related to the philosophical concept of free will. There is no *absolute* free will or responsibility, because all of us are part of the system of interacting entities (human beings interacting in a human society). We can influence others, but are also being influenced by them. *Absolute* free will dogma would imply that we could influence others while ourselves being immune to the influences of others. We cannot influence the behavior of the whole system (which would correspond to the notion of *ultimate* free will and responsibility), but we can modify our own behavior subject to constraining interactions with other entities, other human beings [19]. This latter description represents *ordinary* sense of free will and responsibility and is evidently possible, TOE notwithstanding [19].

Let us comment briefly on experimental insight into the problem of free will. Some recent experiments have shown that unconscious brain activity *precedes* consciously felt intention to act in a particular way [23]. It remains to be seen how free the 'free will' is after all! It is important to point out that TOE and its derivative, sociobiology or evolutionary psychology (EP), indicate only tendencies and potential patterns of human behavior, not predetermined paths. For example, radical changes in human behavior and character could be induced by e.g. genetic or pharmacological manipulations which are not part of Evolutionary process. If EP were to argue that Evolution predetermines paths of human development, it would endow Evolution with a teleological role. This would contradict the very essence of TOE.

Finally, let us suggest that TOE is consistent with general properties of open dynamical systems. The boundary conditions (environmental constraints) induce the formation of order and structure in the system. The energy exchanges with the environment drive the whole process of Evolution inexorably forward. Since Evolution works on populations of organisms, stochastic mechanisms ensure diversity (mutations and copying errors). TOE can be considered a special case of general laws and principles which govern dynamical systems [8].

The discussion above has amply demonstrated how Science on its own i.e. even without Technology can profoundly impact on social values and how easily it gets involved in philosophical disputes. Social and political agendas are often passed off as scientific on the basis of dubious interpretations of scientific results. Such interpretations violate principles of scientific method, have strong value-laden components and should be scrutinized very carefully.

Society's impact on Science

The relationship between Science and values is a reciprocal one. How do social values and priorities affect the way in which Science operates?

Modern societies have overwhelming desire for increased material prosperity and continuous economic growth which exerts societal pressures and constraints on Science. These desires shape funding and research priorities in Science. Other socially driven factors like personal ethical values, financial gain, patriotism or pacifism may determine whether a person is prepared to work on a project with specific military or industrial applications. This brings us to the important question:

Should there be constraints on scientific research in view of the fact that some research may lead to dangerous technologies or that Science needs to produce socially useful outcomes?

The answer is NO, for two very good reasons. Firstly, the increase in scientific knowledge does no harm, ignorance does. What does harm are highly value-laden implications drawn from scientific results and accepted without due scrutiny.

Secondly, we cannot predict future consequences of scientific research at a particular time as the case below demonstrates. The unforeseen or unintended consequences of scientific research may sometimes be beneficial rather than harmful (e.g. discovery of X-rays or laser).

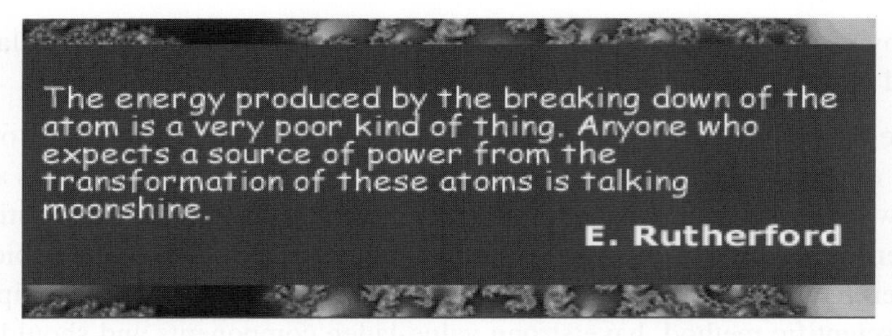

The energy produced by the breaking down of the atom is a very poor kind of thing. Anyone who expects a source of power from the transformation of these atoms is talking moonshine.
E. Rutherford

Rutherford's quotation about the impossibility of harnessing nuclear energy is so much more indicative when we remember that he was one of the pioneers of nuclear physics which made nuclear power generation possible.

The verdict concerning particular technological development should be provisional and open to continuous evaluation. Equally importantly, Society should devise social mechanisms and enhance public awareness of ST issues so as to effectively control technologies once they have emerged.

Science and public policy – climate change [25]

We shall describe the issues related to Science and public policy on one of the best known cases: global warming. The problem of global warming contains two components: scientific and political/economic. Let us describe the scientific component first (see diagram of energy transfers in the color diagram which follows).

- The atmosphere is a mediating layer between Earth's surface and Sun and in this layer the absorbed solar energy circulates. The atmosphere contains gases (water vapour, CO_2, ozone) which absorb some of the solar radiation and thus prevent it from reaching the Earth. The solar radiation (energy) which was absorbed by the Earth surface (in UV region) is subsequently reemitted at longer wavelengths (IR region). Ozone absorbs harmful UV radiation which is desirable, so we should have high stratospheric ozone concentration. However, CO_2 and water absorb IR radiation (heat) which leaves the Earth's surface and thus trap the released energy; this in turn leads to the increase of surface temperature ('greenhouse effect').

- The net energy balance (incoming sunlight vs. outgoing IR radiation) determines the equilibrium temperature at the Earth's surface.
- The temperature range between -20 and 400^{0}C has to be maintained for Life to survive on Earth, so the trapping of certain amount of solar energy is crucial. An increase in average temperature of Earth surface (global warming) may have serious global consequences e.g. changes in crop yields, melting of polar ice caps, rise in the ocean levels etc. The increase in CO_2 emissions due to the use of fossil fuels is considered to be the most important cause of global warming.

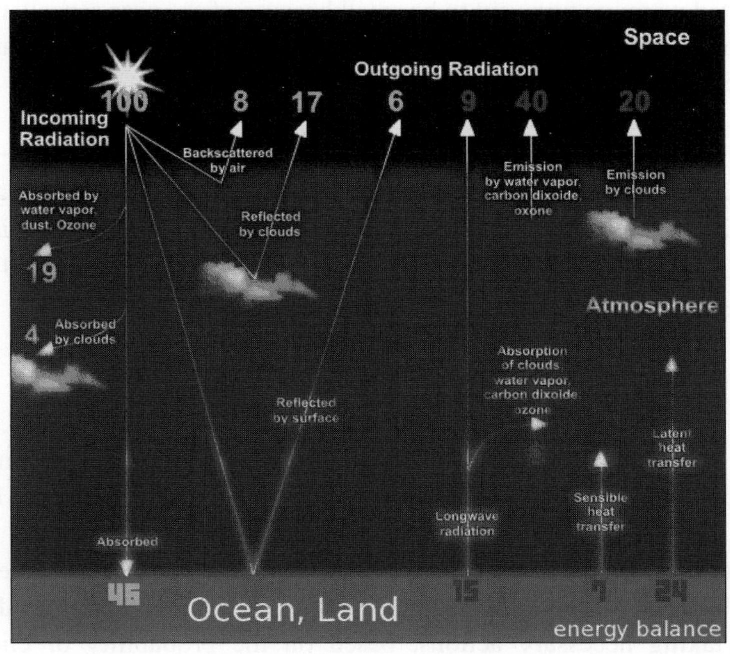

source: http://www.wpclipart.com/energy/informational/energy_balance.png.html

Is global warming prediction true?

This question is a typical scientific problem and can only be answered through scientific research. Governments therefore ask scientists to provide the answer. The subject of scientific research in this case is atmosphere which is a large and very complicated dynamical system. This makes it difficult for scientists to understand all the processes taking place in it and to make reliable predictions. General circulation models are used to make predictions and to simulate future developments in the atmosphere i.e. possible global warming. However, we do not have good data on global temperatures from the past so we cannot test the model's predictions. The existing data about past surface temperatures refer mostly to developed countries; there are no past temperature data for developing world, or the past temperature data for oceans, for the effects of volcanic eruptions etc. To further complicate matters, the net feedback causing the expected warming turns out to be small because of the near cancellation of two large effects. Feedback mechanisms which increase and decrease the surface temperature are shown in the concept map which follows.

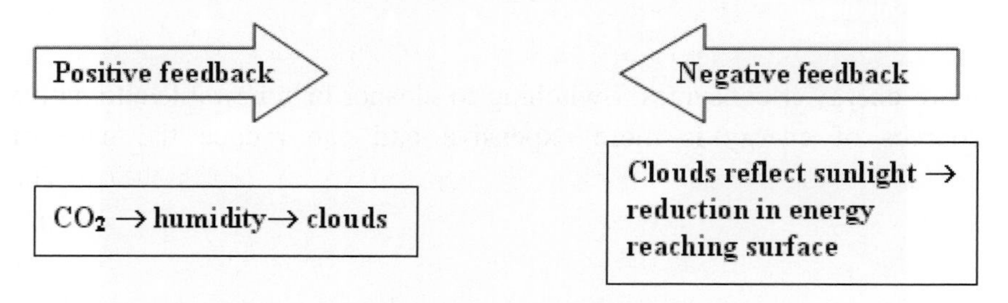

Contributions to greenhouse effect originate from: water 60%, CO_2 26%, ozone 7%, methane and nitrous oxide 7%. The current models predict atmospheric absorption (for clear skies) as being 30 W/m^2 lower than the measured value ('absorption anomaly'). Because of these uncertainties regarding models, Science can only calculate the probability of warming occurring. World nations must therefore decide upon taking necessary actions, based on the probability of climatic events occurring (as predicted by scientific research) and potential damage caused. While this state of affairs is not completely satisfactory it is better than relying on random guesswork or on economic lobbies for guidance. This type of situation often arises when dealing with complicated systems; the answers are given in terms of probability rather than certainty.

The important part of the solution to this problem is development of sustainable and renewable energy technologies which can replace fossil fuel technologies. The development of such technologies would require significant and costly breakthroughs in material science research [25e].

The next important component of the global warming problem is political/economic. Reduction in CO_2 emissions can be achieved either by using modified fossil fuel technologies and by reducing fossil fuel emissions or by the introduction of new sustainable technologies. Either action incurs considerable economic cost which has so far stalled significant global action on climate change. Fossil fuels (coal, oil) are currently the cheapest energy sources, but they emit CO_2 which is one of the causes of global warming. The increase in CO_2 emissions is related to the economic development; the more developed a society/country is the more energy it consumes. Switching to cleaner or renewable/alternative sources of energy is more expensive and can reduce the pace of economic development (which is the sacred cow of modern politics). The intertwined political and economic components are the greatest obstacles to solving the climate change problem as was pointed out recently [25b,c]. The other important obstacle is poor communication between scientists and public on the issue of climate change [25d]. There are several reasons for poor communication:

- Scientists do not consider the problem of global warming sufficiently novel to be interesting as a research topic.
- Scientists who communicate effectively with public may be disadvantaged in their careers or accused of having political agendas if they embrace global warming issue.

- Scientist's career rests largely on peer-reviewed publications and on citations of his work so getting involved in debates about topics which mix science, politics and business is not of great personal interest.

We shall conclude by pointing out the prevalent fallacy regarding Science and matters of public policy. Science should certainly be involved in guiding public policy on global issues, but it cannot do it alone. Global problems require coordinated efforts of scientists, politicians and business corporations. These social groups are quite diverse in their aims and character which implies that solutions to global problems like climate change will not be an easily reached. Let us hope that urgency of the situation and irreversibility of its effects enforces the necessary coordination.

Naturalistic Fallacy: Science can and should decide on matters of public policy!

This version of fallacy is very common and is incorporated in many decisions made by governments or business community. For example, if Science claims that nuclear power is safe, the governments are prepared to build nuclear power stations.

Questions involving 'ought' or 'should' are trans-scientific and have no answer in Science ('naturalistic fallacy'). Scientific knowledge *alone* cannot be used for any public policy decision!

Science, Risk and Evidence [26]

Assessment of risk is as old as human civilization. We know that ancient Babylonians faced this problem and have attempted to develop rational/scientific methods to tackle it. Babylonians had to decide how large their food storage capacity needed to be. If they stored too little food, there will be shortages during poor harvests. If they stored too much food, the food would be wasted. Science is often involved in risk assessment, but this activity suffers from a serious misconception which is related to the artificial dichotomy between expert and non-expert, scientific and non-scientific input. The two cartoons which follow highlight the importance of considering risk in a wider context when performing risk assessment i.e. not relying on scientific evidence alone.

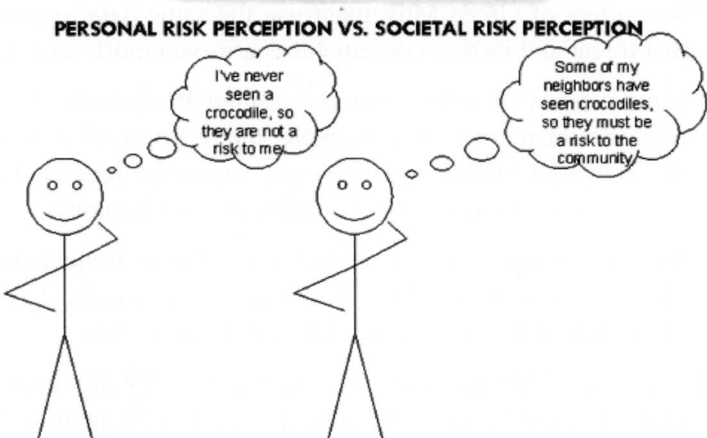

sources: http://www4.gsb.columbia.edu/ipimages/ideasatwork/05-psychology-of-risk.gif
http://crocdoc.ifas.ufl.edu/posters/images/belize_p1/personalrisk.gif

Societal values influence the interpretation of scientific risk assessments. Scientists believe (mistakenly!) that their facts are not value-laden and that their risk assessment is objective ('scientific'). However, this is not true for the following reasons:

- there is rarely a complete and definitive set of data available (e.g. data on atmospheric processes) for risk assessment.
- different scientific models may be used to describe the risk (time frames, significance levels)
- the data used may not be directly relevant (e.g. toxicological studies of new pharmaceuticals/cosmetics are often done on animals rather than on humans, but the harmful doses and other effects may be different for humans)
- scientific data used for risk assessments may contain large uncertainties and consequently risk assessments may give imprecise estimates. The proponents of unrestricted economic growth like to point out that the uncertainties inherent in the atmospheric models and data make the prospect of global warming less threatening than it appears. On the other hand, supporters of Green Movement use the same data to emphasize that threat and its dire consequences are even more severe.
- some aspects of risky situations cannot be quantified (e.g. the role of human error). For example, the response of staff running the industrial facility in emergency cannot be predicted as was the case in industrial accidents in Bhopal or Chernobyl.
- risk probabilities do not reflect the relative frequencies with which events occur. This is because we usually have only incomplete data about the occurrence of past events.

All risks are interpreted through filters that reflect social values. Risks involve many factors: economic, political, social and scientific. This is why risk assessment cannot be left to scientific experts alone because the risk assessment does not comprise only scientific assessment of hazards.

Finally, the important point regarding acceptability of risk is that those who may be adversely affected by the consequences should decide if the risks are acceptable or not.

How should scientific evidence and risk be evaluated?

Let us look at an example of how scientific evidence is evaluated by considering the question whether acupuncture alleviates pain in the lower back (see the concept map which follows).

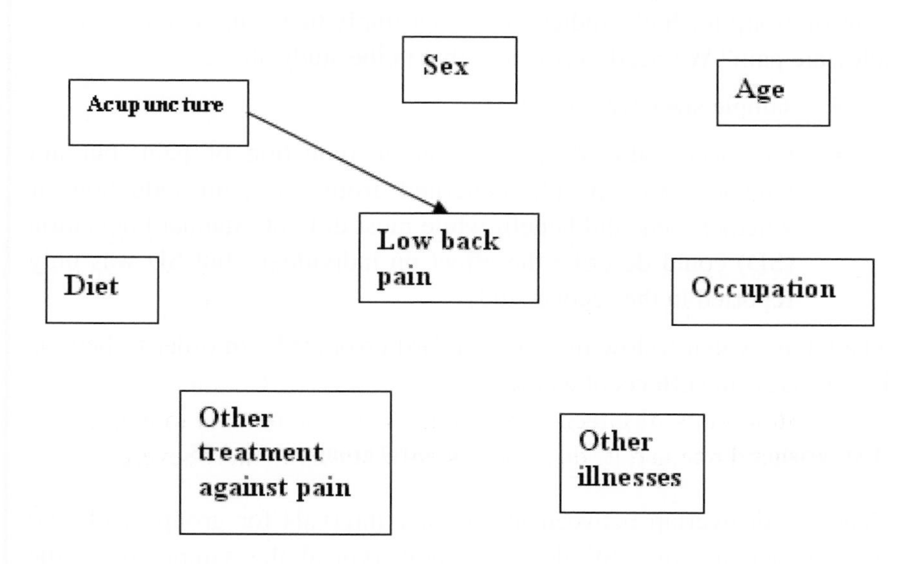

Such study is complex. For instance, continuous data are desirable for dependent variable (DV) of back pain, but how can pain be measured? We can for example, ask patient to select a number on the continuous scale to describe his pain. The randomized control trial (RCT) method was used in two acupuncture studies to account for control variables in the circle of variables above. The numerical results of the acupuncture study are given in the Table below.

	Exp. group	**Exp. Group**	**Control group**	**Control group**
	Size	*Mean (SD)*	*Size*	*Mean (SD)*
Study 1	23	2.7 (no report)	16	4.7 (no report)
Study 2	30	5 (5)	10	6.1 (1.75)

The mean value of pain in the experimental group was lower than in the control group for both studies, so does it imply that acupuncture helps to alleviate pain? We need to consider that in the study above:

- sample sizes were small,
- the mean value describes average reduction of pain, but not whether most people benefited from the pain reduction or whether some did benefit while most did not. Standard deviation (SD) could describe the effect on individuals, but SD was only reported in the second study.

The tables which follow include standard errors (SE) in order to help us better assess the effects of acupuncture.

Exp. group = 4 ± 5	**Control group = 6.1 ± 1.75**

There is an overlap between uncertainty intervals for groups with and without acupuncture. SE describes how typical the sample is of the whole population. Since we need to know how probable it is to benefit from acupuncture, let us consider SE at 68% confidence level.

Exp. group (1SE) 4 ± 0.91	**Control group (SE) = 6.1 ± 0.55**
Pain range 3.1 - 4.9	**5.5-6.6**

At 68% confidence level there is no overlap between the two groups and we can conclude with 68% probability that acupuncture works.

However, what if we want to be 95% certain about the validity of acupuncture treatment?

Exp. group (2SE) - 4 ± 1.8	**Control group (SE) = 6.1 ± 1.1**
Pain range 2.2 - 5.8	**5 - 7.2**

There is an overlap now! The probability that acupuncture works for an individual is therefore between 68% and 95%. This particular study may not be too convincing. It is important therefore to consider as many reported studies regarding acupuncture as possible. If all the reported studies point in the same direction, i.e. if they exhibit small bias in favor of acupuncture, then this may enhance the case for usefulness of acupuncture.

Scientific evidence/data is only one of three factors which may influence final risk assessment, decision and data interpretation. Also important are:

- social values (e.g. should we favour young or middle aged people in medical treatment)
- available resources and cost of medical treatment

Let us look at one more example to illustrate the issues involved in risk assessment and data interpretation. A long distance truck driver is prescribed a drug which has been scientifically proven to cause drowsiness as a side-effect. Will the driver:

- not take the drug,
- look for alternative, less effective medication
- adjust his lifestyle (change jobs) and take the drug?

The concept diagrams which follow illustrate that risk assessment and evidence analysis have layer-like structure. Scientific data and information are certainly the starting point in the decision making process. However, before the final (best?) conclusion can be drawn or action taken we need to transverse other layers which represent other important contributing factors.

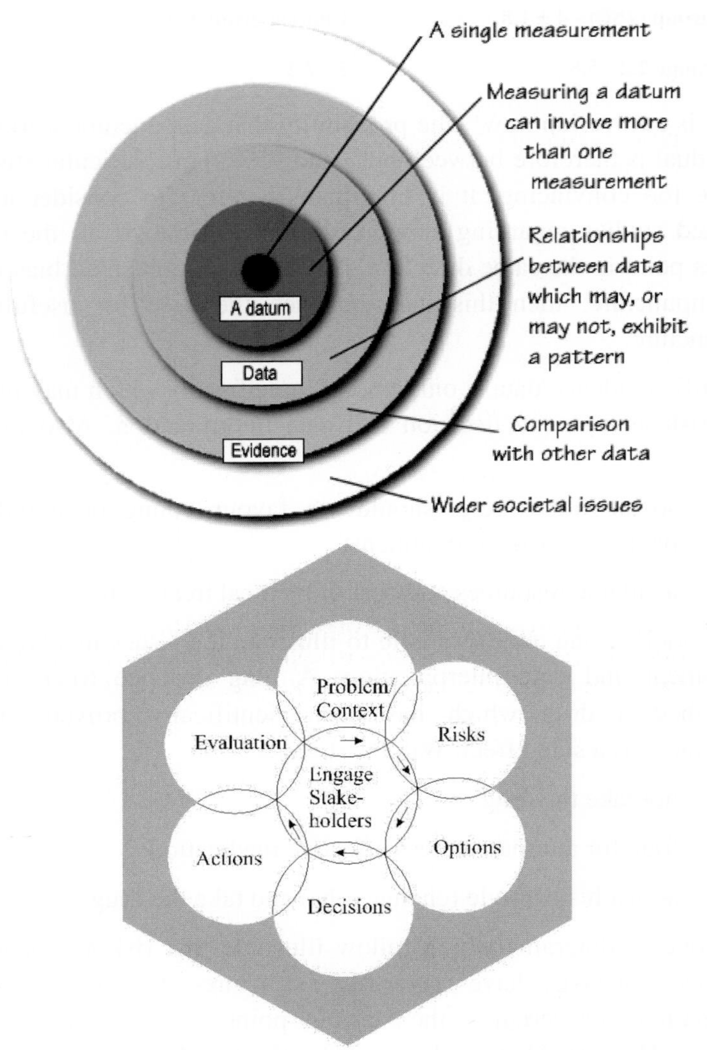

source: http://www.dur.ac.uk/rosalyn.roberts/Evidence/cofev.htm
source: http://biosafety.abc.hu/unep/per3.jpg

Reflection:

Under what social conditions (if any) may costs of scientific progress outweigh its benefits?

In the ancient World (see Chapter 4) which were greater: the costs or the benefits of Science?

Hints:

Consider what governs and controls scientific activity.
Consider the basis of economy in the ancient World.

Bibliography for Chapter 3:

1. a) Mackay, A.L. (1991) *A Dictionary of Scientific Quotations* (IOP Publishing, UK).
   b) Gaither, C.C., Cavazos-Gaither, A.E. (2000) *Scientifically Speaking* (IOP Publishing, UK).
2. Stevenson, L., Byerly, H. (2000) *The Many Faces of Science* (Westview Press, 2nd ed. USA).
3. Bridgstock, M., Burch, D., Forge, J., Laurent, J., Lowe, I. (1998) *Technology and Society, An Introduction* (Cambridge University Press, UK).
4. a) Haste, H. (1997). Myths, monsters, and morality. Understanding 'antiscience' and the media message, Interdiscip.Sci.Rev. 22, pp.114-120.
   b) Boulter, D. (1999). Public perception of science and associated general issues for the scientist, Phytochemistry, 50, pp.1-7.
5. Burrow, J.W. (2000) The Crisis of Reason. European Thought 1848-1914 (Yale University Press, USA).
6. Silver, L.M. (1998), It's not the meek who will inherit, The Times Higher Education Supplement (16 Jan, UK) pp.16.
7. Pennisi, E. (2005). How did cooperative behaviour evolve?, Science, 309, pp.93. [1 Jul issue].
8. a) Schneider, E.D., Sagan, D. (2006) *Into the Cool. Energy Flow, Thermodynamics and Life* (The University of Chicago Press, USA).
   b) Csermely, P. (2009) *Weak Links*, (Springer-Verlag, Germany).
9. Patzig, G. (2002). Can moral norms be rationally justified?, Angew.Chem.Int.Ed.Engl. 41, pp.3353-3358.
10. Porter, H. (2009) The right to offend, *Guardian Newspaper*, (23 Sept, UK).

11 a) Nowak, M.A. (2006) *Evolutionary Dynamics* (Harvard University Press, USA).
b) Nowak, M.A. (2006). Five rules for the evolution of cooperation, Science, 314, pp.1560-1563. [8 Dec issue].
12. Tovee, M.J., Maisey, D.S., Emery, J.L., Cornelissen, P.L. (1999). Visual cues to female physical attractiveness, Proc.Roy.Soc.London Ser.B, 266, pp.211-218.
13. Munoz-Rubio, J. (2002). Sociobiology and human nature, Interdiscip.Sci.Rev. 27, pp. 131-141.
14. Reich, E.S. (2009). The rise and fall of a physics fraudster, Phys.World, (May issue), pp.24-29.
15. McCurry, J. (2006) Disgrace, Observer Newspaper, (1 Jan issue, UK).
16. Hermanowicz, J.C. (2009). A physicist's life-cycle, Phys.World (Sept issue) pp.42-44.
17. Ašperger, S. (2008). Faust's deliberations, Croat.Chem.Acta 81, pp.525-527.
18. Djerassi, C., Pinner, D. (2003) Newton's Darkness. Two Dramatic Views (Imperial College Press, UK).
19. Richards, J.R. (2000) Human Nature after Darwin. A Philosophical Introduction (Routledge, UK).
20. Alexander, D. (2001) Rebuilding the Matrix (Lion Publishing, UK).
21. Freeman, G.R. (1990) Kinetics of nonhomogeneous processes in human society: Unethical behaviour and societal chaos, Can.J.Phys. 68, pp.794-798.
22. Jenkins, S. (2009) The war on drugs is immoral idiocy. We need the courage of Argentina. *Guardian Newspaper*, (3 Sept issue, UK).
23. Schuster, P. (2009). Free will, information, quantum mechanics, and biology: It pays to distinguish different forms of free choice and information, Complexity, 15, pp.8-10 (Oct issue).
24 a) Strenger, C. (2009) Dawkins is wrong about believers, *Guardian Newspaper*, (4 May issue, UK)
b) Vernon, M. (2009) God, Dawkins and tragic humanism, *Guardian Newspaper*, (11 June issue, UK).
25 a) Kirk, J.T.O. (2007) *Science and Certainty*, (CSIRO, Australia).
b) Browne, L. (2009). Challenges in tackling climate change, Phys.World, (Oct issue) pp.20-21.
c) Reynolds, C. (2009). Regulatory burden, Chem.World (Dec issue) pp.45.
d) Romm, J. (2009). Publicize or perish, Phys.World (Oct issue), pp.22-23.
e) Crabtree, G., Sarrao, J. (2009). The road to sustainability, Phys.World, (Oct issue) pp.24-30.
26 a) Mehta, M.D. (2002). The social construction of science, risk and expertise, Can.Chem.News, (March issue) pp.28-30.
b) Gott, R., Duggan, S. (2003) *Understanding and U*sing Scientific Evidence Sage Publications, UK).

Chapter 4
Science and History

(All quotations in this chapter are taken from ref. [1])

Science is a historical phenomenon, a product of civilization which came into being during certain phase in human history.

The reason for studying historical aspects of Science is to illustrate by historical examples how Science evolved and what Science's main characteristics are. History of Science may also give us some clue about future developments pertaining to Science. The consensus in contemporary societies is that Science has become indispensable and that future of human civilization depends on it. Science currently appears to be the only human activity about which such broad consensus exists.

A few specific questions regarding historical development of Science are given below:

Where did Science originate from? Why did it evolve?
Why did Modern Science begin 16th century AD in W. Europe?
What were/are the geographical centres of scientific activity?
Why there?
What is the future of Science?

The discussion of the above questions can help even scientists to better understand their own activity.

The only thing wrong with scientists is that they don't understand science. They don't know where their own institutions came from, what forces shaped them and are still shaping them and they are wedded to an anti-historical way of thinking which threatens to deter them from ever finding out.

E. Larrabee

The concept map which follows shows how inter-related the development of scientific ideas was and how this inter-relatedness lead to the evolution of modern Science. One may conclude that no national group or geographical region can claim the honor of "inventing" modern Science.

"Development of Western Science is based on two great achievements; the invention of the formal logical system (in Euclidean geometry) by the Greek philosophers, and the discovery of the possibility to find out causal relationship by systematic experiment (renaissance). In my opinion one has not to be astonished that the Chinese sages have not made these steps. The astonishing thing is that these discoveries were made at all."

A. Einstein

In fact, as Einstein has pointed out in his quotation the development of Science is such a unique phenomenon that it could only have occurred through broad cooperation, exchange and development of ideas which originated in different regions and cultures. The exchange of physical devices and know-how (technologies) followed a similar pattern. The concept map on history of ancient and medieval science has a 'funnel shape' which points to Europe as the original source of modern Science. Note also how the neck of the 'funnel' in the map starts to broaden again in modern times. This broadening is directly related to globalization processes which favour broad dissemination of knowledge.

The history of ancient and medieval science

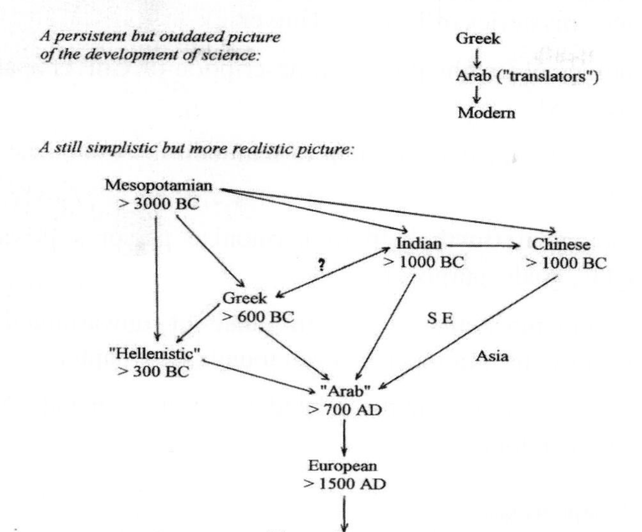

source: http://fritsstaal.googlepages.com/history.jpg

The same concept map reveals intellectual interdependence and multicultural contributions which shaped the development of modern Science. The next concept map below shows how different scientific disciplines evolved. In historic terms Science is one of many different attempts by mankind to explain and dominate Environment so that we can survive as individuals and as a species.

Myths represent the first attempt by mankind to develop a belief system or a World view. Myths provided explanation, ordering and even prediction of events thus giving some much needed feeling of comfort and security to human race. 'Myth' comes from the Greek word *mythos,* which means 'word' and it literally means an authoritative account of the subject whose value is not to be questioned.

Myths gave birth to Art and Science (see the concept map which follows).

Myths had various roles in the past including:

- metaphysical (myth maintains respect and humility towards ultimate mysteries of Life and Universe);
- cosmological (myths provide a description of Universe and how it functions)
- social (myths justify and help to maintain the established social order)
- psychological (myth help to harmonize person's private life, thoughts, goals, purpose)

We need to remember that myths in guise of unwarranted public perceptions or misconception persist even today (see Chapter 1).

Examples of better known myths related to Science include Myth of Prometheus and Pandora.

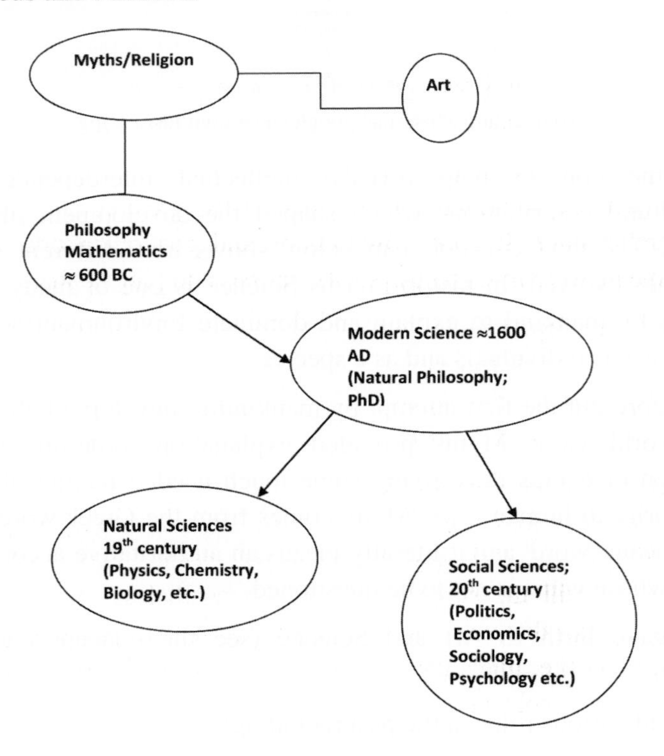

Greek philosophers began (in 5th century BC) replacing mythological explanations by descriptions of facts which could be observed, proved and debated. This was a momentous and necessary step towards the development of modern Science. However, it was only the first step in the long series of efforts towards developing science; it did not represent Science in the modern sense of the word.

Aristotle described logical deduction as a systematic method of thinking. Deduction explains why things happen and uses cause and effect principle. This allowed explanations to be made by reference only to observable and rationally accessible entities, unlike myths which postulated the existence of supernatural entities and effects.

Technology evolved before Science. It started from everyday activities and needs for sustaining Society. In fact, the first technological activities were making of tools and implements for agriculture, construction, war, clothes and cosmetics. Technology *did not* require special social pre-conditions to emerge; it was driven by everyday necessity. Technological developments were carried out by artisans, craftsmen, builders.

On the other hand Science *did* require social pre-conditions before it could emerge. Science is driven by curiosity and the desire to discover and understand Nature. It seeks abstractions and generalizations. Scientific developments were initiated by priests, scribes and philosophers.

Until 19th century, Science and Technology were separated from each other due to social forces and economic conditions. Technology and Science are still different activities even though in modern Society they strongly reinforce and cooperate with each other.

One of the main questions in the history of Science is what factors (social, physical) were important in the development of Science [5]. The sequel to this question is which (if any) factors were the most important. All these questions are still the subject of debate. Nonetheless, current historical research suggests that we should not use modern concepts, perceptions and points of view (i.e. hindsight) as a starting point for historical analysis of Science. It is simple and tempting to try and find pieces of historical evidence (events, ideas) which bear similarity to our current notions and subsequently label/interpret these pieces using modern terminology/theories. Scientific developments in the past need to be understood on their own terms i.e. as the expressions of individual and social goals of their own times and driven by social forces which existed at that point in time. The importance of two sets of factors regarding scientific developments is however established. One set of factors is related to the physical environment of a particular society or civilization (e.g. geographical location, available natural resources, natural landscape features which govern communication and transport especially towards neighbours, etc.). The other set of factors is the fabric of particular society itself: its beliefs, ethics, economics and political organization. Some authors have suggested that the first set of factors determines the second [2,6]. While the extent of this determining influence is open to discussion, there is no doubt that physical environment does strongly (but not exclusively) influence the nature of a particular society and through it the scientific activity which is carried out in such society. In our subsequent discussion we shall use the 'linear/sequential model' which is (somewhat simplified) summarized in the concept map below. This simple model can explain many developments even though it emphasizes the role physical environment at the expense of other factors like historical exchanges which took place between different societies and the human psychological factors. The existence of independent, parallel developments in different societies and cultural cross-fertilization between societies suggests the need to use a more comprehensive 'parallel model' in history of Science. We shall introduce these other factors into the discussion when necessary i.e. when the linear model is not satisfactory.

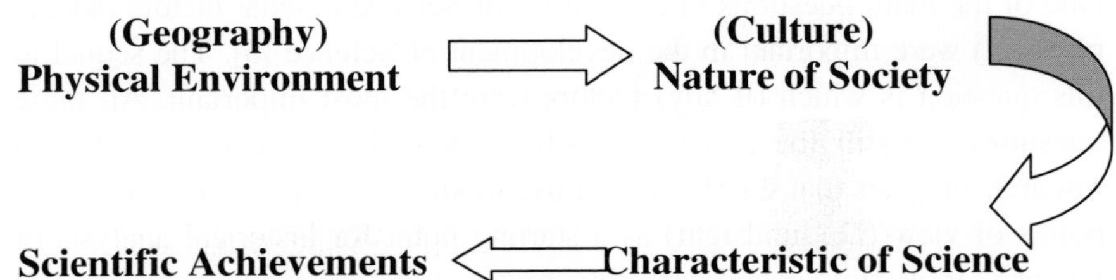

Science is a complex activity (see Einstein's quotation) whose existence depends on many pre-conditions and is only possible once a Society reached certain level of economic and cultural development. Around 6000 BC first civilizations emerged independently in different parts of the World thus establishing the environments in which some scientific activities could be performed. The pre-conditions for scientific activity included:

- establishment of agricultural surpluses, division of labour (manual, intellectual)
- development of writing and counting
- growth of cities (e.g. Babylon, Xian)
- construction of monumental architecture (pyramids, Great Wall)
- metal-working

Two traditions were particularly important in the early stages of Science. The first is called 'bureaucratic tradition' (or 'tradition of professionals') and the second 'tradition of amateurs'.

A. Bureaucratic tradition [2,3]

This tradition emerged and existed in hydraulic civilizations e.g. Egypt, Mesopotamia, China (2000BC-1850 AD) which depended on irrigation agriculture like those shown in the following map.

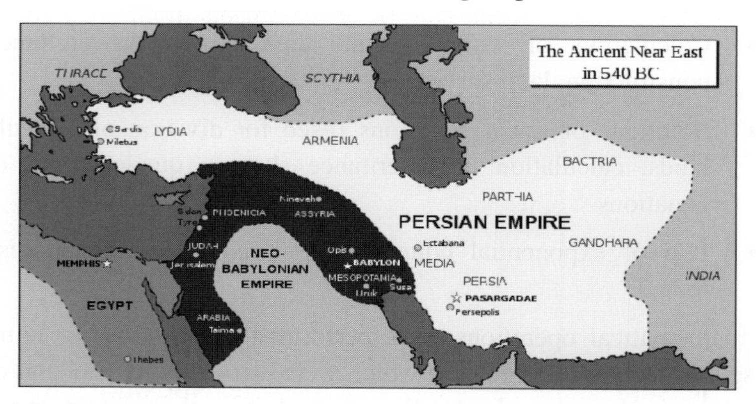

source: http://commons.wikimedia.org/wiki/File:Ancient_near_east_540_bc.svg

In this tradition knowledge and learning were supported and organized by state and temple authorities mostly for practical purposes. 'Scientists' were civil servants (scribes, priests) they were educated at special schools and acquired specialized skills: doctors, accountants, astronomers, engineers or teachers. They were often anonymous because of their civil service status.

Subjects of scientific studies were limited to practical purposes (calendar, architectural construction, business transactions). Knowledge was acquired through experience, and was recorded and used in the form of lists, rather than as theorems or generalizations. 'Science' at that time did not attempt to understand Universe, Society or human beings. The World view was based on magical and religious practices. There was no independent study of Nature as such e.g. unrelated to Gods and magic. Heavens were considered Divine and magical.

No distinction was made between scientific, rational and pseudo-scientific, magic knowledge. Usefulness was the only criterion. This is why astronomy/ astrology and chemistry/ alchemy were equally supported and accepted in these civilizations.

Examples of scientific achievements in the 'bureaucratic tradition':

Mathematics

- Calculation of volumes and areas (used in architectural construction, land surveying)
- Solutions of linear equations (used for division of agricultural lands, calculation of inheritance shares), quadratic and cubic equations
- Use of exponential equations (for calculation of interest on loans).

The mathematical operations were performed using tables of numbers and sets of recipes (procedures) which were arrived at by trial and error. There was no abstract understanding of numbers or their manipulations.

Astronomy

- Setting up of calendar systems for agricultural purposes (planting, harvesting, dating land contracts etc.).
- Babylonian astronomers (Babylon was in modern Iraq) manipulated data related to solstices, solar & lunar cycles, they made predictions of solar eclipses, rising, setting and visibility of planets (e.g. Venus).
- Some Babylonian astronomical records date from 2000 BC. Continuous observations were made from 747 BC.

Babylonians performed the first recorded scientific research. The topic of that research was to answer the question:

What is the interval between subsequent full moons: 29 or 30 days?

This is a many-variable problem; the solution to the problem depends on the season of the Year, apparent Sun-Moon distance in the sky and long term lunar cycles. The Babylonian astronomer-priests used observation, mathematical analysis and modeling and have created tables which gave accurate predictions of the intervals sought.

Civil and military engineering

- construction of roads, aqueducts, irrigation canals, pyramids, catapults, fortifications
- building of pneumatic machines and automata.

B. Greek tradition or 'tradition of amateurs'

This tradition arose and flourished in mercantile city-states of Greece and the Eastern Mediterranean from 600 BC till 500 AD which are shown on the geographical map which follows.

This unique set of events in the Greek World evolved due to distinct geography and society prevailing in the Greek geographical area. In

Greece the agrarian base was poor due to poor soil and there was no possibility of building large scale irrigation systems. Such environment favored development of overseas commerce, political decentralization and generally outward looking societies which were not deterred by geographic barriers from their communication with neighbors.

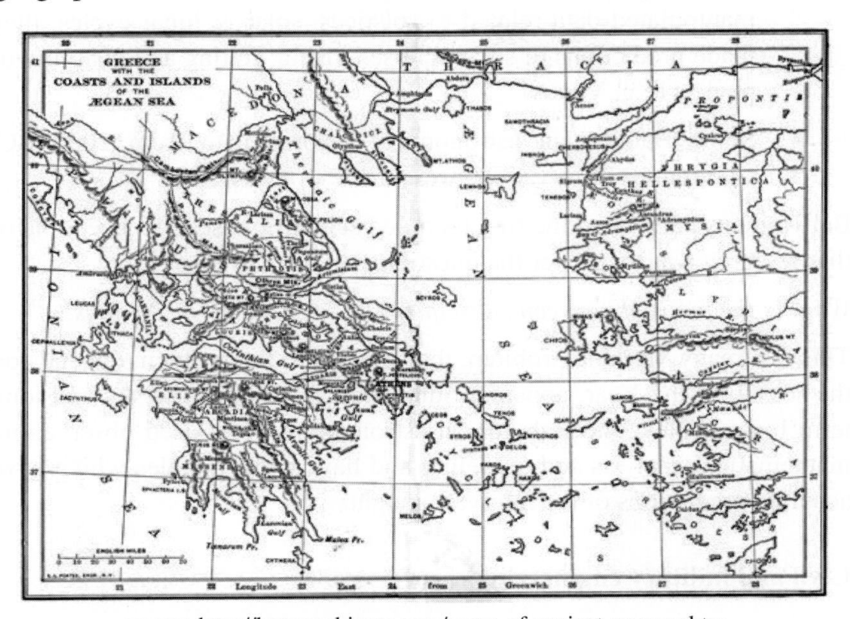

source: http://karenswhimsy.com/maps-of-ancient-greece.shtm

The characteristics of scientific activity in this 'tradition' are:

- Science was performed by recognizable individuals with no state support and outside of state institutions.
- Scientific investigations were driven by curiosity rather than by usefulness.
- Knowledge was organized on the basis of abstract, general theories and was obtained by logic (rational argument). Nature was deemed comprehensible to human intellect which

gave Science its original name "natural philosophy", that is philosophical inquiry directed at Nature.

- Occult and magic learning was displaced by rational and philosophical arguments.
- Subjects of investigations were abstract and general. For example, what substance is the World made of? How do we know about the World and how it evolved into its present state?

Examples of achievements in the 'Greek tradition' include:

Mathematics

- Development of the concept of number as the central unifying principle of Nature by Pythagorean philosophers. Pythagoreans believed almost religiously in mathematical order and representation which alone would unlock mysteries of Universe.
- Introduction of the notion of rigorous proof by deductive reasoning.

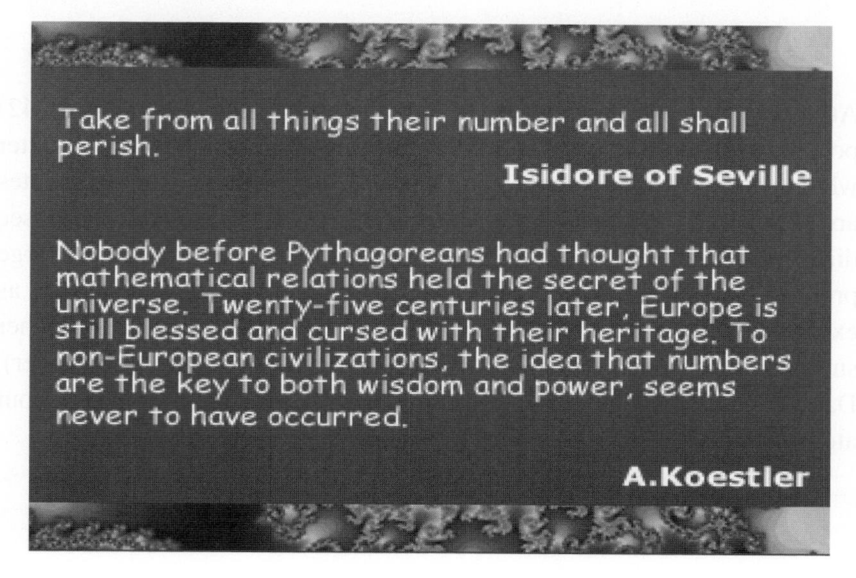

Take from all things their number and all shall perish.
Isidore of Seville
Nobody before Pythagoreans had thought that mathematical relations held the secret of the universe. Twenty-five centuries later, Europe is still blessed and cursed with their heritage. To non-European civilizations, the idea that numbers are the key to both wisdom and power, seems never to have occurred.
A.Koestler

To compare the two traditions let us recall that Babylonians and Egyptians were familiar with Pythagoras theorem which they deduced by trial-and-error and used in practice. However, unlike the Greeks, they did not see the need to formally prove the theorem.

Physics

In physics (to use the modern term) there was development of general, abstract theories which attempted to explain the whole of Nature without recourse to the supernatural forces or ideas, but only to the regular processes observed in Nature.

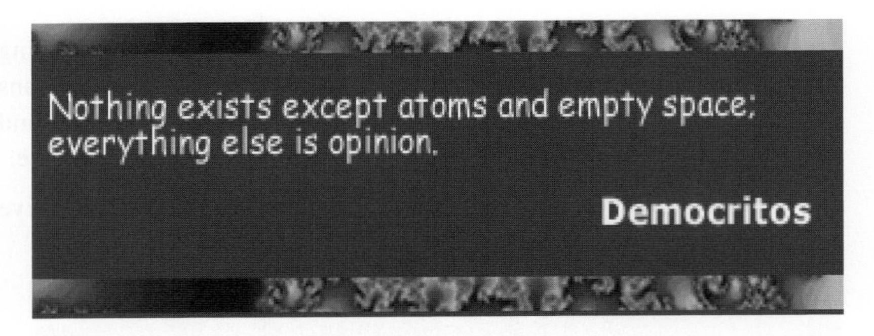

Nothing exists except atoms and empty space;
everything else is opinion.
Democritos

Atomistic theories of Democritos, Leucippus and Heraclites (Figure 42) postulated the existence of atoms as irreducible constituents of all matter which can form innumerable combinations (aggregates). The aggregates are held/locked together by protuberances on the surfaces of atoms (see diagram below). Some atoms have (according to Greek atomists) large protuberances which make the corresponding bulk of matter harder as exemplified by iron atoms sketched below. Other atoms have smoother surfaces which enable them to slide over each other with ease (water). Dalton's atomic theory in 19th century maintained similar ideas about atoms equipped with hooks.

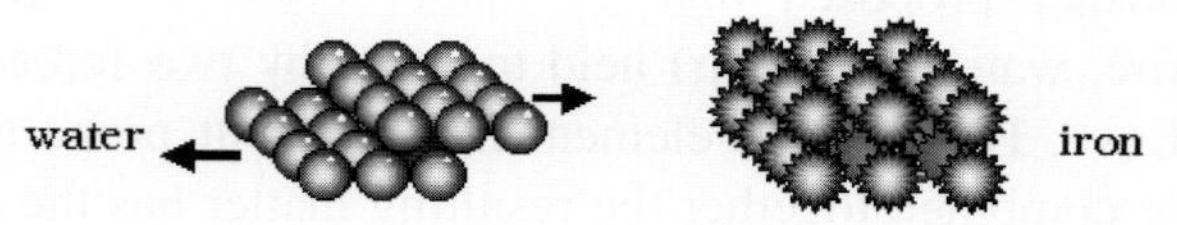

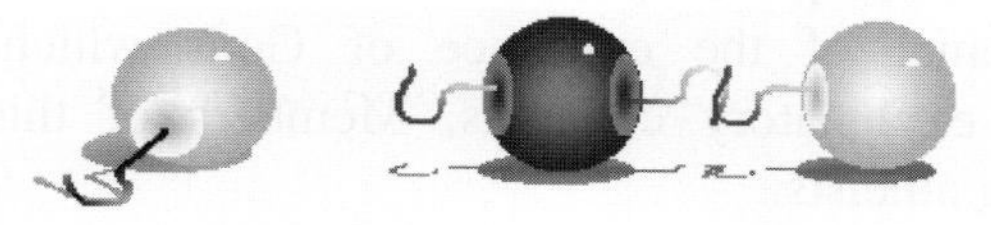

source: https://reich-chemistry.wikispaces.com/file/view/atom_prop.gif

Fig. 42.

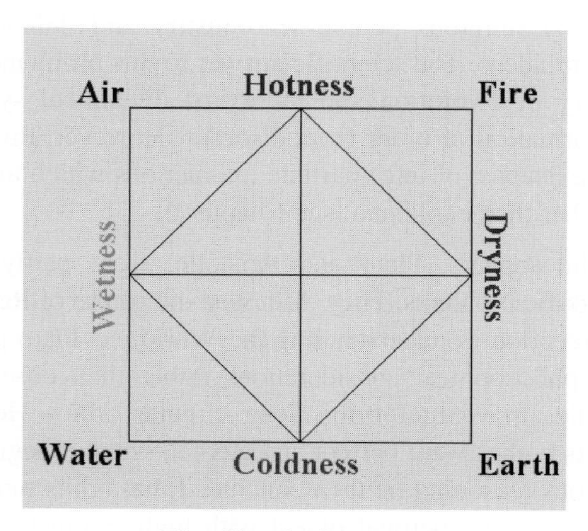

The Greek elements.

source: http://abyss.uoregon.edu/~js/images/greek_elements.gif

Fig. 43.

Heraclites is supposed to have said: *'The World consists of atoms and void. The atoms are in perpetual motion.'* These ideas explained the diversity of material world (atoms are different and enter into various combinations with each other) and perpetual change in the World. Atomists further proposed that all matter consists of atoms of four elements (fire, water, earth, air) held together by two forces (attractive, repelling; Love, Hate). Each element has different properties so when elements are combined together the resulting matter has the combination of the two properties (Figure 43).

Soul according to this philosophical school is also built of atoms which lead to the denial of the existence of Gods which thus became superfluous as explanatory concepts. Members of this school were accused of being atheists.

The main problem with atomistic theory of the World was its inability (at the time) to explain how can randomly moving atoms create order, patterns and objects which we see around us. The same objection is raised today by religious people who justify and defend their Faith against atheistic ideas. The scientific answer to this problem is embedded in The Theory of Evolution and theory of dynamical systems which explain the formation of order from disorder. However, this explanation relies on the existence of inter-particle interactions which are themselves postulated rather than explained (see Chapter 2).

The great philosophers, Plato and Aristotle, were partly involved in studying scientific problems. They discussed the nature of Reality, the role of human perception in understanding the World etc. Plato postulated, on the basis of philosophical considerations rather than observations, that heavenly bodies move uniformly along circular orbits. He argued that Heaven and its bodies were perfect and eternal, without beginning or end. Using analogous reasoning he then concluded that orbits must be circular because circle is a geometrical object with high symmetry (perfection?) and without beginning or end. This is a good example of dangers involved in the use of analogy (see Chapter 2 and Figure 44).

Another famous Greek philosopher Aristotle was also active in trying to answer scientific questions related to the theory of motion, classification of organisms and biological processes. His scientific work included observations, but not experiment.

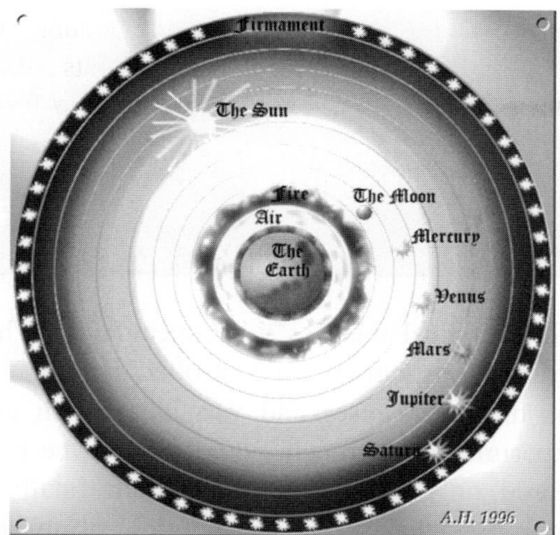

Plato's model of the solar system - one of the first cosmological models
source: http://abyss.uoregon.edu/~js/images/greek_cosmos.jpg

Fig. 44.

The following is an example of Aristotle's scientific reasoning obtained by deductive logic and supported by observation, but not by experiment:

> *Vacuum is impossible! Why? In vacuum a body would experience no resistance to its motion and would thus acquire infinite velocity. The infinite velocity is impossible because it would allow the body to be present in two places at the same time.*

This conclusion is based on the logical proof called *'reductio ad absurdum'* i.e. the transformation of premises until logical contradiction/impossibility is reached. However, the argument above comprises logical *fallacy of hidden assumption* which turns out to be wrong. The hidden assumption is that body needs continuous application of force to keep it moving. Aristotle has correctly observed that all bodies require constant input of force (contact with the mover) for a motion to proceed, but this is due to frictional forces which oppose the motion. The reliance on observation alone could not always establish scientific correct facts and relationships as Russell's pun below indicates.

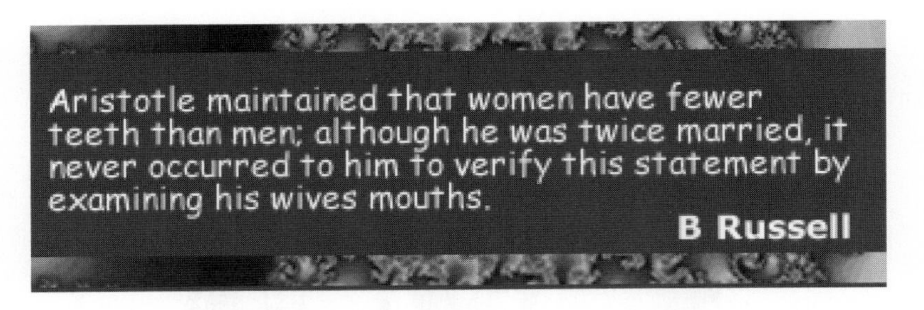

Aristotle maintained that women have fewer teeth than men; although he was twice married, it never occurred to him to verify this statement by examining his wives mouths.
B Russell

Some Greek philosophical theories were speculative. They were based on logic and observations, but not on controlled experiments which could identify various multiple causes and variables which govern complex phenomena. Experimental approach would resolve the problem, but this approach appeared only around 1600 AD and finally gave birth to modern Science. Nonetheless, these theories firmly established the practice of rational argument which is necessary for Modern Science.

Combined tradition

The combination of both these traditions did exist briefly in the Ancient World, but the political, economic and social forces brought 'New World Order' to Europe after 500 AD and destroyed much of classical knowledge. Nonetheless, the characteristics of combined tradition are interesting and represent a foretaste of future developments towards modern Science.

In the city of Alexandria (in modern Egypt) the '*Museum*' and '*Library*' (Figure 45) were established where scholars were provided with state facilities (library, lecture halls, accommodation) to support independent research programs. The programs were active in the period of about 280 BC-500 AD and the knowledge accumulated was recorded on half a million scrolls of papyrus. Alexandria was a great learning centre of the ancient world. However, towards the end of the era, social conditions changed and the last stipendiary of the Museum, Hypatia was killed in a riot by proponents of anti-intellectual movements (around 415AD).

A scroll room of the Library of Alexandria.
source: http://www.wpclipart.com/world_history/arts_science/Library_of_Alexandria.png.html

Fig. 45.

Some examples of scientific achievements in the combined tradition are follow:

Mathematics

- development of axiomatic geometry by Euclides
- study of conic sections (parabola, hyperbola etc.) by Apollonius

Physics

- proposition of the heliocentric system by Aristarchus
- introduction of elaborate mathematical model of the Solar system by Ptolemy

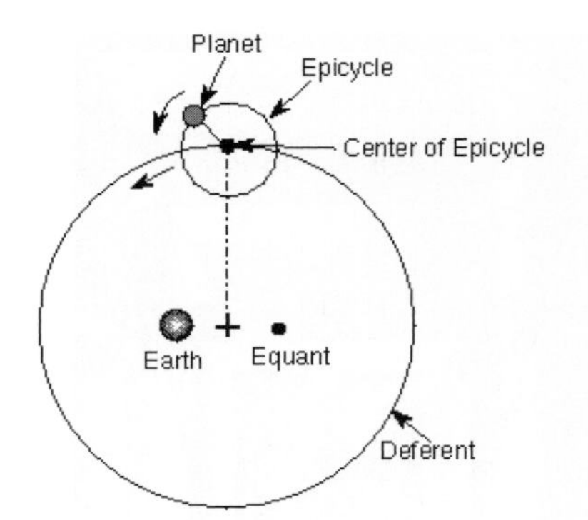

Center of epicycle moves counter-clockwise on deferent and epicycle moves counterclockwise. Epicycle speed is uniform with respect to equant. The combined motion is shown at right.

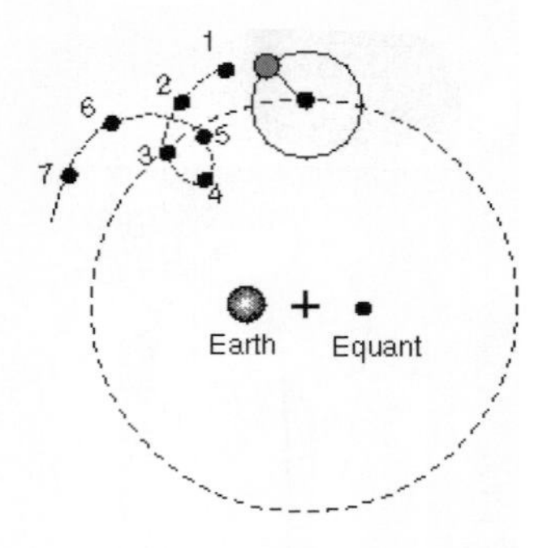

Deferent motion is in direction of point 1 to 7 but planet's epicycle carries it on cycloid path (points 1 through 7) so that from points 3 through 5 the planet moves backward (retrograde).

source: http://www.astro.virginia.edu/class/oconnell/astr121/guide06.html

The two diagrams which follow demonstrate how the explanations of the motion of celestial objects were arrived at by Ptolemy. The main effort of scientists was to develop a theory which would reconcile the notion of immovable Earth as the centre of Universe with the observed planetary movements (retrograde motion). The notion that Earth is stationary was arrived at by religious, mythical and philosophical considerations and was considered to be beyond dispute. Special orbits ('epicycles') were designed for the purpose of explanation. This is an example of the influence of social factors on scientific research which persists till this day. For, example the modern notion of human rights and values influences modern scientific research (e.g. stem cells research and genetic engineering).

The geocentric model shown in the previous two diagrams is more complicated than the heliocentric model depicted in the diagram which follows and by using the principle of Ockham's razor (see Chapter 2) we would select the latter model. However, heliocentric model postulated that Earth is not stationary and does not reside in the centre of Universe which was 'politically' unacceptable for a long time. Here is an example of 'political correctness' extraordinaire!

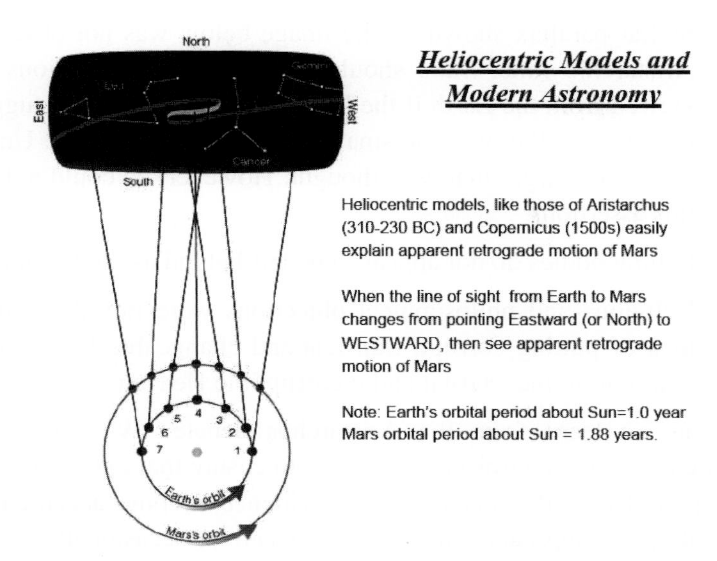

North
East
West
Cancer
South
Heliocentric Models and Modern Astronomy
Heliocentric models, like those of Aristarchus (310-230 BC) and Copernicus (1500s) easily explain apparent retrograde motion of Mars
When the line of sight from Earth to Mars changes from pointing Eastward (or North) to WESTWARD, then see apparent retrograde motion of Mars
Note: Earth's orbital period about Sun=1.0 year Mars orbital period about Sun = 1.88 years.
5 4 3
6 2
7 1
Earth's orbit
Mars's orbit

source: http://www.astro.virginia.edu/class/oconnell/astr121/guide06.html

The heliocentric vs. geocentric model debate

This debate is a nice illustration of how scientific reasoning proceeds and we shall explore it in some detail.

Aristarchus's proposed heliocentric model which was able to explain simply (satisfying Ockham's razor principle) the daily circuit of heaven and path of the Sun around the zodiac. The heliocentric model was fundamentally correct, but was not accepted during his time. The main objections to heliocentric model were a mixture of scientific arguments and philosophical speculations/assumptions:

- Bodies not fixed to the Earth's surface would fly off it if Earth did rotate. Birds fly with equal ease in all directions and they should not if the Earth is spinning.
- Aristotle's notion of heavy objects falling 'naturally' towards the centre of the Universe where they become stationary (reach equilibrium) was contrary to the principles of heliocentric system.

- Stellar parallax shown on the image below was not observed at Aristarch's time. Stars should change relative positions when viewed from the Earth if the Earth rotates. Aristarchus suggested that the parallax was too small to be observed because Universe was much larger than was thought. However, he could not prove this assertion.
- Falling bodies do not appear to be left behind as the Earth moves
- Religious and philosophical objections were raised against the idea of putting corrupt, transient and changeable Earth into the realm of divine, eternal and incorruptible Heaven.

What can we learn from the Aristarchus debate? We can see that scientific ideas are scrutinized sceptically to ensure that only those which are supported by sufficient empirical evidence become accepted. This methodological approach may (paradoxically) occasionally lead to correct explanations being rejected in the short term, but it ensures that they are accepted in the long term and that Science can surely progress.

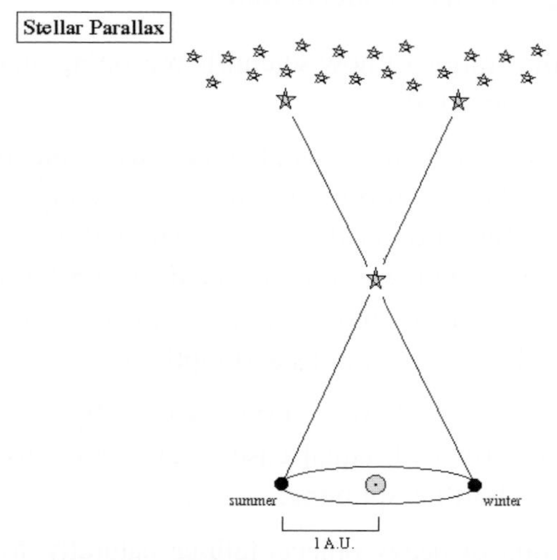

parallax is the apparent change of position of a nearby star with respect to background stars due to the motion of the Earth around the Sun

source: http://www.sciencebuddies.org/mentoring/project_ideas/Astro_img004.gif

Mechanics

- Studies in mechanics and fluids by Archimedes and Hero lead to the invention of various mechanical devices, automata and the discovery of buoyancy principle. The diagram below illustrates the device which is similar to modern steam turbine and converts heat into mechanical energy.

source: http://www.grc.nasa.gov/WWW/K-12/rocket/TRCRocket/IMAGES/History1.gif

Geography

- calculation of the circumference of Earth by Eratosthenes

After 500 AD, the Science in Ancient Greco-Roman World (European region) collapsed due to political and social turmoil, but the scientific tradition continued to be preserved, enriched and independently developed by Islamic, Chinese, Indian civilizations. However, modern Science was not born in any of these civilizations. Why? This is one of the most interesting, frequently debated and unanswered questions in the history of Science [7-9]. We shall compare the scientific developments in Europe and China next. We do not imply that Islamic, Indian or other civilizations did not make contributions towards the endeavor which eventually became modern Science. We focus on a typical example to highlight difficulties inherent in debates about why modern Science evolved where it did and when it did. These debates often touch on sensitive issues related to national pride and identity. Science in modern context is considered to be the most influential and advanced of human activities so any 'claim' which a particular national/ethnic group makes about being the 'originator' of modern Science is bound to ruffle feathers and recall historical grievances. Nonetheless, rational explanation should proceed without reference to emotional arguments. The three scientific traditions, West Eurasian (including ancient Mesopotamian, Greek and Islamic), Chinese and Indian have similarities and differences in character and emphasis [9]. The differences are more instructive and we shall mention a few of them. Eurasian tradition was characterized by emphasis on Nature, reductionism and discrete (non-continuous) mode of development. This mode is reflected in oscillations in the level of sophistication applied to scientific problem solving. Chinese tradition was characterized by applied nature of scientific endeavors, experimentation (trial-and-error) and by the influence of historical and organic (holistic) outlook on Science and the World. Indian tradition was characterized by a high level of theoretical abstraction and formal analysis (reductionism) in solving scientific problems as well as the application of scientific theory to human society and ethics [9]. In spite of these differences the three traditions achieved similar level of development until 16th century when the European part of Eurasian tradition overtook developments in all other regions. The three traditions represent three geographical centers of scientific activity. With European ascendancy the centre of activity remained for a while in Western Europe only to be dispersed again in the second half of the 20th century when scientific efforts became globalized.

Examples of Scientific Development: China, Europe

We shall follow the 'sequential model' in our analysis. China has large, fertile, alluvial planes (Yangtze, Yellow River) which are surrounded (sealed off) in the North, West, South West by geographical barriers: deserts, mountains, steppes. Chinese was the most isolated of Old World civilizations and its geography impeded extensive contacts with W. Asia and Europe for many years. China has low fragmentation of its regions compared to high fragmentation exhibited by European geographical zone (see the maps of two regions which follow). The fragmentations represent rivers, mountain barriers, lakes, coastlines etc.

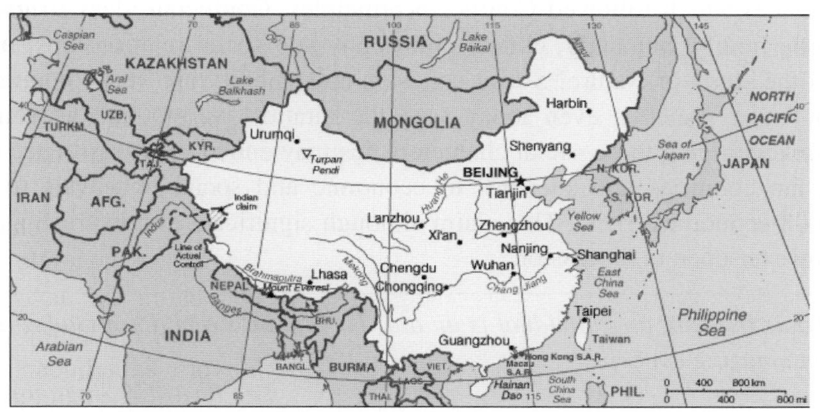

source: http://www.pdclipart.org/displayimage.php?album=71&pos=86

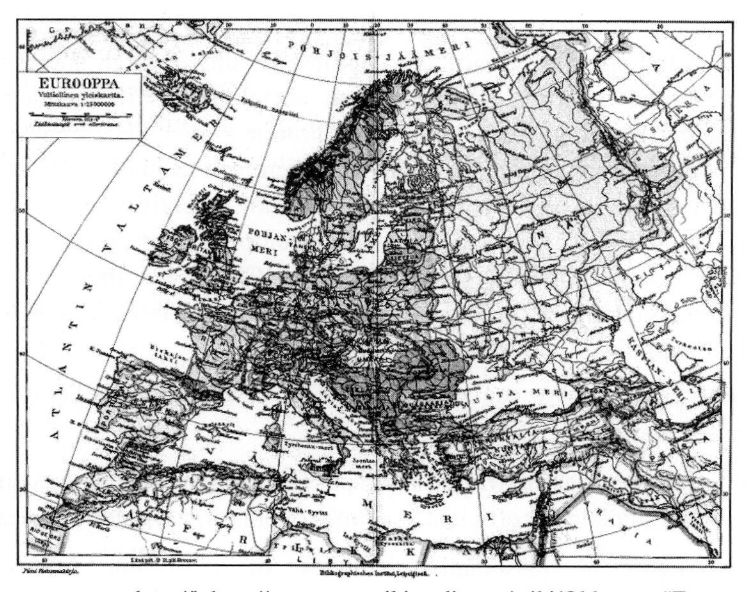

source: http://hthttp://commons.wikimedia.org/wiki/Old_maps#Europe

Such geographical (environmental) conditions favoured the development of densely populated, highly centralized state in China with cultural and social unity and continuity [6a]. The state power was concentrated in the hands of the Emperor and state bureaucracy (mandarinate → civil service) which followed Confucian principles. Confucian ideas focused upon family, humanity, ethical philosophy and social relationships, **not** on the study of Nature! Social cohesion and stability are very important in Asian societies even today (see the citation below) and as some thinkers argue, these social characteristics may enhance the resilience of Asian countries and societies to economic and social upheavals (e.g. 2008 economic crisis). This claim although significant and interesting is however open to discussion.

> *To us in Asia, an individual is an ant. To you, he's a child of God. It is an amazing concept.*
>
> *Lee KY (MM Singapore)*

The social attitude exhibited by China is typical of most hydraulic civilizations. The vast and powerful civil service was organized and trained in the spirit of Confucian classics, it enjoyed high social status and prevented the rise of independent social bodies like colleges of independent learning, trade guilds, autonomous cities. The emphasis of most studies was on practical aims and therefore technology flourished [7] and continuously evolved. There was no 'scientific revolution' in China i.e. there was no accelerated development of Science. Many scientific investigations were performed in China, but there was little organized, coherent effort to develop systematic understanding of Nature via abstract and general scientific theories.

Possible explanations for the absence of "Scientific Revolution" in China:

- Complexities of written and spoken language mandarin which is less suitable for efficient scientific expression and communication
- Analogical (associative) modes of reasoning relying on metaphors

- There was no concept of the "Law of Nature" (holistic view of Nature). This concept originated in societies with monotheistic religions which assumed that the World was rationally ordered by being the product of single God-Creator as the citation below exemplifies.

> *"The book of nature, which we have to read, is written by the finger of God."*
>
> *M. Faraday*

- The social feeling of cultural superiority due to its long unbroken tradition. This has made Chinese less willing to absorb cultural achievements from outside of their own civilization.
- State control of intellectual life and low status of merchant/business classes in ancient China
- Great influence of Confucian & Taoist philosophies on personal outlook on Life and Nature
- State support steered the scientific activity primarily towards the acquisition of useful knowledge.

Achievements of Chinese tradition in technological/scientific fields [7]

- Technology of paper manufacture
- Hydraulic engineering (Grand canal)
- Manufacture of pottery (porcelain)
- Silk manufacture
- Iron metallurgy (use of high temperatures for iron ore smelting)
- Invention of gunpowder
- Invention of magnetic compass

China is a good example showing how the development of Technology preceded and was independent of Science. Scientific and technological developments were often controlled by environmental/geographical factors.

At the end of 15th century scientific developments in Chinese tradition slowed down in relative pace compared to W. Europe. The table below compares the characteristics of two scientific traditions.

W. Europe	China
Rainfall based agriculture, forest and mountain barriers, highly indented coastline	Irrigation based agriculture, no major mountain barriers within vast dry planes Poorly indented coastline
Local political control	Central, state control
Political, cultural diversity based on competition between nation-states	Cultural homogeneity
Autonomous universities	State directed education and examination systems
Outward looking countries/societies engaged in trade and geographical exploration (1492 AD American continent was discovered)	Inward looking civilization (1433 AD abolishment of ocean shipping on the orders of Ming emperor)

It needs to be noted at this point that non-existence of *scientific revolution* in China has been disputed by e.g. Sivin [8] who claimed that the revolution did in fact take place in 18th century China, but that it did not have social consequences which we can expect to accompany such revolution. Even the notion of 'scientific revolution' in Europe has been questioned [4]. Modern historians point out several reasons for doubting that a truly revolutionary break with the past occurred in 16-17th century. Some activities in the 17th century were not fully compatible with modern notion of Science. There is a surprising continuity between knowledge generation in the Middle Ages and 17th century, there was no single scientific method (see Chapter 1) so what could the 'scientific revolution' be all about? Nonetheless, we shall retain the term 'scientific revolution' which though not a perfect description does bring across the awareness which 17th century scientists ('natural philosophers' actually) had about being engaged in truly novel range of activities and obtaining truly remarkable results.

European civilization was based on the rainfall dependent agriculture which requires moist, heavy, clay soils. The crop yields in this type of agriculture are less predictable and abundant than in irrigation based agricultures. This is why European civilization flourished later than other great civilizations. The clay soil also requires the use of heavy, iron plough, horses, horseshoes and similar agricultural implements. On the other hand, there was no need for extensive irrigation schemes which favour centralized state organization, no great public works and no state sponsorship of Science and Technology in Europe.

It is therefore not surprising that there are no buildings comparable to Great Wall of China or Pyramids of Egypt in Europe! See earlier citation by Einstein regarding development of modern Science.

Scientific Revolution in W. Europe (16th-18th century)

(preconditions, content, achievements, consequences)

Factors which facilitated the birth of Modern Science:

- Improved agricultural production, increased population density, urbanization, improved economic conditions (development of market economy)
- Relatively easy geographical access to Islamic civilization which preserved and improved ancient scientific knowledge. W. Europe **was not** the seat of one of the old civilizations, but was geographically close to the Mediterranean intellectual legacy (Greece & Rome) which was preserved in the Islamic world! The closeness to ancient sources of learning was *necessary, but not sufficient* condition for the development of modern Science! Islamic civilizations of the Middle East were in possession of ancient scientific traditions, but due to social and religious factors were unable to continue with the development of modern Science.

- The establishment of Universities in large European urban centres. Universities (from 1100AD on) were a uniquely European "invention". University was a corporation consisting of Masters and Students. It was a *legally constituted corporation* which had independence from local and state authorities. The University's autonomy was possible because of weak central State/Church control. The weak control allowed learning of pure Science to be established in the curriculum. University was a "marketplace" for competing ideas and opened to anyone who paid the fees which were charged directly by Master-scholars to their students. This idea of knowledge as a commodity is nicely described in the quotation below. It is interesting that the status of knowledge as a commodity (market driven and valued) is widespread today. This status of modern knowledge may be one of the constraints on the future development of Science as shall be discussed later. Nonetheless, the University may still have important role in the future globalized World as the institution dedicated to exchanging and developing new ideas and methods needed to solve global problems [10]. University is possibly (because of its traditional intellectual independence and the spirit of free inquiry) a better place to foster growth of knowledge than economic or political institutions like industrial corporations or government research centres which are often subservient to the interests of small but influential social groups.

> *"Only a very small part of any ordinary person's knowledge has been the produce of his own observation or reflection, all the rest has been purchased in the same manner as his shoes or his stockings, from those whose business is to make up and prepare for market that particular species of goods."*
>
> *A. Smith*

- Limited centralization in W. Europe was reflected in the formation of groups of competing nation-states. This limited centralization was achieved through "military revolution" (i.e.

> necessities of production of firearms, fortifications, ocean-going navies) and not through demands of irrigation agriculture. Ironically "military revolution" had negligible direct input from Science. It was made possible by practical experience, skills, use of rule-of-thumb principles and bold experimentation by craftsmen, engineers, navigators.

Nonetheless, "military revolution" has stimulated technological development, colonialism and global conquest. We again note the analogy with modern times when many scientific developments were prompted by developments in modern weaponry. The social impetus for change in 16-18th century was enormous and it ushered in "Renaissance". In such atmosphere the 'scientific revolution' was readily initiated.

Philosophical/metaphysical ideas in Christian Europe were often underpinned by "theism" i.e. the belief in the order in Nature which was created by God and which is intelligible to humans. Such metaphysical basis of ideas was lacking in other civilizations which did not adopt monotheism. Of course monotheism is only one of the influential factors which could have facilitated *scientific revolution*. For example, the Islamic world did not develop modern Science as W. Europe did in spite of its monotheistic religious beliefs. Science has long since left this theological cradle behind, but it is interesting that theology and Religion may have been catalysts for *scientific revolution*. Ironically, the development of Science itself has reduced the importance of Religion in many developed countries. Scientific revolution was a complex social and cultural process; it was a series of events leading to changes in worldview over time.

> *"Science developed only when men refrained from asking general questions such as: What is matter made of? How was the Universe created? What is the essence of Life? Instead they asked limited questions such as: How does an object fall? How does water flow in a tube? Thus, in place of asking general questions and receiving limited*

> *answers, they asked limited questions and found general answers. It remains a great miracle, that this process succeeded, and that the answerable questions became gradually more and more universal."*
>
> *V. F. Weisskopf*

Achievements of the 'scientific revolution'

- Change in the understanding of Universe and human body (intellectual transformation), a revision of understandings of Hellenistic & Eastern scientific traditions.
- Emergence of new scientific methods: controlled experiment, mathematical modelling.
- New ideas about the potential of Science to change the World and about the social usefulness of Science.
- Changes in the social role and institutional organization of Science

The general concept of the exploitation of natural resources for the purpose of improvement of human and social conditions was uniquely European and evolved together with the development of democratic societies. Political ideology and ST developments were the engines of progress, first in W. Europe, than throughout the World till the present day. Knowledge is power! The environmental constraints/problems accompanying such unlimited development were not recognized at that time.

The table below summarizes some key scientific developments in order to illustrate that *scientific revolution* was not a single localized event (in time or space), but a series of semi-independent intellectual advances ('tremors') spanning time and space.

1540
publication of Copernicus' *De revolutionibus orbium coelestium* ("On the Revolutions of the Celestial Spheres")

1632
publication of Galilei's *Dialogo sopra i due massimi sistemi del mondo, tolemaico e copernicano* ("Dialogue on the Two Main World Systems, Ptolemaic and Copernican")

1679
formulation of Newton's Laws

1684
formulation of the Universal Law of Gravitation

1661
formulation of Boyle-Mariotte Law on the relationship between pressure and volume of gases

1774
discovery of oxygen

1789
publication of Lavoisier's *Traité élémentaire de chimie* ("Elementary Treatise of Chemistry")

1860
First International Chemical Congress

1868 - 1870
publication of Mendeleev's *"Principles of Chemistry"*

1735
publication of Linnaeus' *Systema Naturae* ("The System of Nature")

1809
publication of Lamarck's *Philosophie zoologique* ("Zoological Philosophy")

1859
publication of Darwin's *"On the origin of Species by Means of Natural Selection, or the Preservation of Favoured Races in the Struggle"*

Further examples:

The scope of achievements listed below demonstrates that Scientific Revolution was not a series of unconnected (fortuitous) discoveries of limited practical use, but rather a continuous, systematic effort towards discovery and understanding of Man and Nature.

Astronomy and mechanics

Heliocentric model of Solar system proposed quantitative laws of planetary motion devised on the basis of observation ***(Copernicus, Kepler)***. The model was much more efficient in explaining astronomical observations than the geocentric model. Its adoption is an example of the application of Ockham's principle.

Controlled experiments on the motion of bodies were performed by ***Galileo*** to understand mechanical motion. Studies of Moon and other planets by telescope gave experimental support to heliocentric system (*Galileo's discovery of the phases of Venus)*

Generalization of the laws of motion and discovery of universal gravitation was made by ***Newton***. Newton's mathematical model of mechanics unified and simplified our understanding of Nature (celestial & terrestrial domains). It has established that the same mechanical laws (regularities) hold throughout Universe. This was very different from ancient thought which considered Heaven & Earth to follow different laws and have different characters.

Medicine & optics

Experimental study of human anatomy as the basis of medicine by ***Vesalius***.

Discovery of "tubes" in the human body by ***Eustachi, Fallopius***.

Discovery of the circulation of blood by ***Harvey***.

Chemistry

Definition of chemical element and construction of air-pump for studying gases by ***Boyle***. Discovery and measurement of gas pressures using barometer by ***Torricelli.***

Broader consequences of scientific revolution:

Scientific revolution was part of very important events/processes in world history. It helped to make countries of W. Europe (WE) politically and economically dominant in the World until 1940 when USA used the legacy of modern ST imported from Europe to forge ahead, untouched by WWII devastation and political/social/economic turmoil in Europe. However, 'scientific revolution' was only one of the factors which contributed to the rapid growth of Western civilization and their societies.

New relationship between Science and Technology

Modern Science remained separated from Technology intellectually and socially until 19th century and during that time Technology had greater impact on Science than vice versa! (e.g. through the invention of microscope, telescope, chronometer).

Industrial revolution, i.e. change from agriculture to manufacturing as the primary economic activity was NOT initiated by Science, but by ecological pressures on Society. The inventors of technological processes for iron manufacture and steam engine (*Watt, Stevenson, Newcomen*) were engineers and craftsmen, not scientists. However, from 19th century onwards Science and Technology became inextricably linked for the first time in human history. This would not have been possible if scientific activity remained confined to metaphysical theorizing.

> *"Science owes more to the steam engine than the steam engine owes to science."*
>
> *J.B. Conant*

Societal support for Modern Science

In the 16th century scientific activity was conducted by individual scientists at their own initiative/expense, as it was in antiquity. Universities mainly transmitted the existing knowledge (were involved in teaching), but some research was also done under the auspices of the University.

Under the influence of developments in the Society at large, Science underwent structural changes. Initially, scientific investigations were performed only at Universities, later scientific societies with moderate state support were formed to coordinate scientific activities (e.g. publish journals, sponsor prizes & expeditions, organize surveys). In the 19th century specialization/professionalization occurred in Science and specialized scientific societies were founded to reflect this. Also, in the 19th century strong State and business community support for Science began; Science became a recognized profession. Since then Universities took the twin roles of teaching and research in almost equal measure. The dates of foundation of Scientific societies ("academies") given below illustrate the development of institutionalized Science and increased involvement of Society in nurturing the development of modern Science.

Foundation dates of scientific academies	
1660	Royal Society of London for the Promotion of Natural Knowledge
1666	Académie des Sciences (France)
1700	Akademie der Wissenschaften (Germany)
1724	Akademiya Nauk (Russia)
1743	American Philosophical Society (USA)
1780	American Academy of Arts and Sciences (USA)

Modern experimental science is very expensive and requires societal support, but such support comes at a price which often involves the loss of independence in choosing the subject of inquiry as well as the method.

> *"The real accomplishment of modern science and technology consists in taking ordinary men, informing them narrowly and deeply and then, through appropriate organization, arranging to have their knowledge combined with that of other specialized but equally ordinary men. This dispenses with the need for genius. The resulting performance, though less inspiring, is far more predictable."* J.K. Galbraith

The future of Science

Science is a historical phenomenon. Its fortunes have risen and fallen throughout history. Science has not run out of problems to solve as the recently published list of fundamental unsolved scientific problems demonstrates [11]. It is ironic that some scientists (see citation below) felt that there is nothing new to be done, but they were quickly proven wrong. They should have known better!

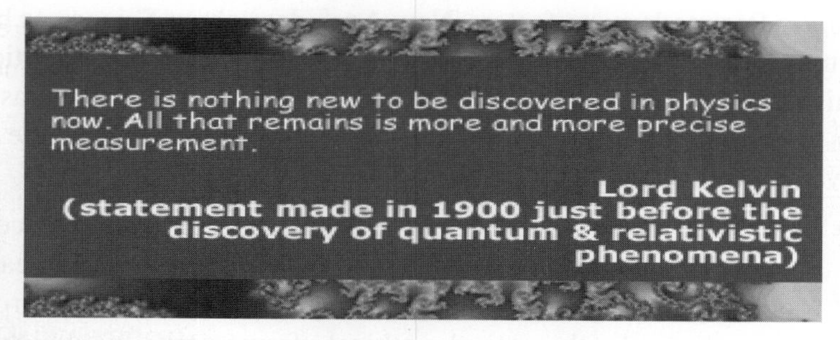

There is nothing new to be discovered in physics now. All that remains is more and more precise measurement.
Lord Kelvin
(statement made in 1900 just before the discovery of quantum & relativistic phenomena)

There is no shortage of new problems or challenges for Science. More to the point, many challenges may be as yet unknown. For example, for scientists of the Victorian era the composition of human genome was not a research problem which they could contemplate, because genetic structure of the cell was then unknown.

Potential causes of scientific decline

It is interesting to consider potential causes of decline in Science as a human activity. The causes include:

- decline in social importance of Science, as well as the lack of scientific career opportunities. This seems unlikely due to extremely high level of dependence of modern societies on Science and Technology; especially for solving pressing problems (overpopulation, aging population, environmental degradation)
- separation between Science and Technology. For, example ancient Greeks did not develop sophisticated technology in spite of their success in fundamental Science, because their economy was based on cheap slave labour. The separation is unlikely to occur in the future for the reasons given above and because Technology cannot function well without Science.
- spread of religious or other fundamentalisms and superstitious practices
- decline in the appreciation of value of scientific knowledge, decline in trust in Science and spread of general scepticism towards intellectual pursuits.

Example: Postmodernism fosters 'New Age' ideas about Science as being a form of constructed knowledge which is valid only within a particular linguistic/social community and having no claim to universal assent. Such trivialization of scientific knowledge may undermine the motivation to investigate physical world.

- public concern about negative/dehumanizing consequences of scientific and technological development (environment, weapons development, genetic manipulation). Since the loss of broader metaphysical, theoretical, ethical framework in which ST operates, ST may become increasingly prone to abuse by special interest groups in Society.
- imposition of free market mechanisms onto scientific practice (for example, Einstein would be unlikely to obtain research funding for his work today)

Some potential causes of decline of Science are discussed in more detail below.

Religious fundamentalism

For the purposes of our discussion the 'fundamentalist view' refers to an exclusive, fixed view of the World which automatically considers different views to be wrong without due rational scrutiny. Proponents of fundamentalist views are not only impervious to rational arguments, but are prepared to exert physical force to impose their views on others. Fundamentalists justify their stand by belief rather than reason.

source:
http://ginacobb.typepad.com/gina_cobb/images/september_11_attack_3.jpg

Fig. 46.

The images in Figure 46 indicate potential destructive consequences for global Society (and Science of course) of the application of extreme religious ideas to World affairs. Religion (especially its monotheistic variety) claims to provide universal view of the World and so does Science. Since the two views have in many cases diametrically opposite notions about Nature and Society, conflicts are bound to arise.

We have mentioned in Chapters 1-2 that neither Science nor Religion has a monopoly on human knowledge. However, extrapolations are often made from precepts of these two forms of knowledge which can create tensions. The economically developed, largely secular societies are skeptical or dismissive about religious claims and invoke the results of Science to support their skepticism. This skepticism which is understandable for people without religious convictions may lead to tensions with Societies in which religious tenets are regarded as indispensible part of their social fabric (see the quotation which follows). The claims of universal validity put forward in religious and scientific descriptions of the World, coupled with their fundamental differences represent a constant source of potential conflict. This happens despite all public protestations about intellectual and religious tolerance. In connection with this religious-secular dichotomy it is interesting to note that Richards [12] in her careful and convincing analysis of social, ethical and personal implications of the Theory of Evolution, concluded that there is an unbridgeable gap between materialist and non-materialist views regarding the future of human civilization. The religious fundamentalism may (but does not have to) 'spill over' into political fundamentalism in societies where religious views are the integral part of their social consciousness. The current concern about terrorism is an example of this spill-over. It would be interesting to see if economic development will reduce religion inspired fundamentalist tendencies (as many in developed countries hope). Fundamentalism in general, is often a sign and instinctive reaction against unsettling social changes and challenges; it is driven by suppressed psychological fears and doubts and by craving for social stability at any cost. Science demonstrates that individuals and societies as open dynamical systems must necessarily be subject to profound, irreversible changes throughout history. We do however have some control over how these changes will happen and which direction they will take. This comment is not the 'glorification' of change for its own sake, but a reminder of the nature of reality.

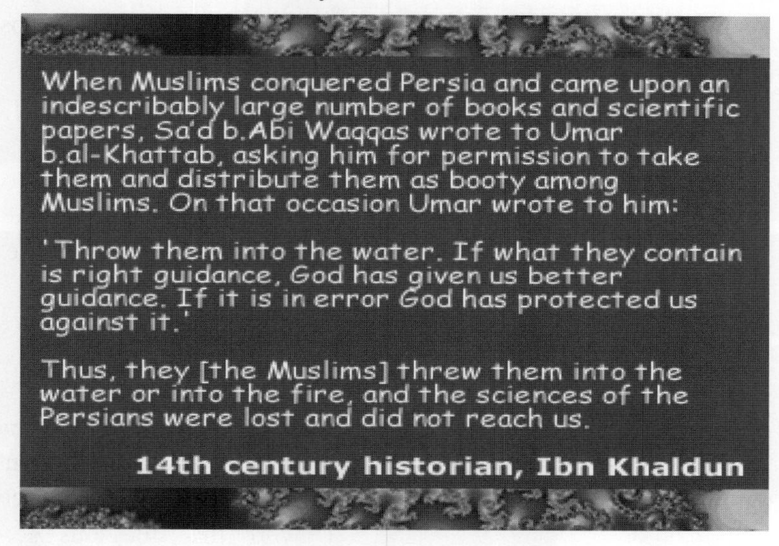

When Muslims conquered Persia and came upon an indescribably large number of books and scientific papers, Sa'd b.Abi Waqqas wrote to Umar b.al-Khattab, asking him for permission to take them and distribute them as booty among Muslims. On that occasion Umar wrote to him:
'Throw them into the water. If what they contain is right guidance, God has given us better guidance. If it is in error God has protected us against it.'
Thus, they [the Muslims] threw them into the water or into the fire, and the sciences of the Persians were lost and did not reach us.
14th century historian, Ibn Khaldun

Political fundamentalism

Religious fundamentalism often, but not always (as USA shows) appears in less economically developed societies. However, developed countries are not immune from fundamentalist instincts either. There are several historical examples of political fundamentalism in the developed countries. Destruction/censoring of information and persecution of intellectuals are often symptoms of Society's fear of new developments and profound changes. Examples of political fundamentalism are given below and are illustrated by Fig 47 and the citation which follows:

- the burning/banning of books and persecution of intellectuals in Nazi Germany, Stalin's Soviet Union and Maoist China
- "Monkey trial" in USA in 1936 regarding the teaching of Theory of Evolution.

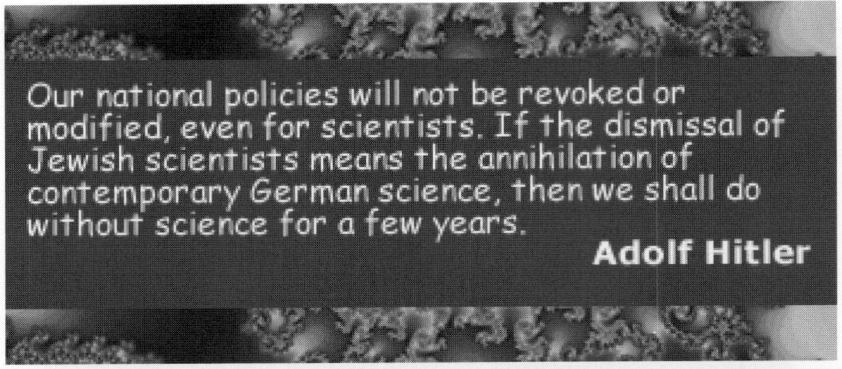
Our national policies will not be revoked or modified, even for scientists. If the dismissal of Jewish scientists means the annihilation of contemporary German science, then we shall do without science for a few years.
Adolf Hitler

The burning of 'unsuitable' books in Nazi Germany

source: http://isurvived.org/Pictures_Isurvived/Book-burning.gif

Fig. 47.

Economic fundamentalism

It is the notion that Science must be subjugated to managerial practices and market laws. Science rests on two important pillars (Laws of Nature AND human Society) and therefore cannot function successfully if it is forced to follow only societal considerations (human instincts, fashions, political and economic agendas etc.). This is because such agendas are not always within the bounds of possible or generally beneficial developments. Unfettered global economic development was initiated on the basis political and economic ideologies and assumptions about 'human happiness' without due considerations for interconnectedness making the World integrated political, economic and ecological system. We are part of Nature, we depend on it and there **ARE** limits to what we can achieve in pursuing our social agendas aiming at 'better life'.

One of the great values of Science is in delivering a 'reality therapy' which warns us of possible overconfidence or blind dogmatism of market economics. Of course, the usefulness of this 'reality therapy' applies to other fields of human endeavor too e.g. to religious or political dogmatism. This argument does not suggest that ST should not be useful for Society, but that ST is not useful as a market commodity only!

The tenets of economic fundamentalism pertaining to Science are further expounded by citations of prominent scientists and the editorial comment in the journal *Nature* (see the following box).

...Science has lost her freedom. She has become a productive force. She has become rich but she has become enslaved and part of her is veiled in secrecy.

P.L. Kapitsa

We are raising a generation of young men who will not look at any scientific project which does not have millions of dollars invested in it... We are for the first time finding a scientific career well paid and attractive to a large number of our best young go-getters. The trouble is that scientific work of the first quality is seldom done by go-getters, and that the dilution of the intellectual environment makes it progressively harder for the individual worker with any ideas to get a hearing...The degradation of the position of the scientist as an independent worker and thinker to that of morally irresponsible stooge in a science-factory has proceeded even more rapidly and devastatingly than I expected.

N. Wiener

The scientific enterprise is full of experts on specialist areas but woefully short of people with a unified world view. This state of affairs can only inhibit progress and could threaten political and financial support for research.

Nature commentary

source: http://www.wpclipart.com/famous/Gates_conquest.png.html

Economic fundamentalism (as mentioned in Chapter 1) aims to encompass not just Science, but all of Society. For example, one of the assumptions of this view is that religious fundamentalism and Religion itself shall inevitably be phased out with the increased material prosperity at the global level. In order to achieve this prosperity it is necessary to boost economic development as the first priority, or so the economic fundamentalism claims. At this point we recall the issues hinted at by Snow in '*The Two Cultures*' (see Chapter 2). Current political and business elites have little Science awareness since their education background mostly comprises Arts, Law or Business studies. This may contribute to the rise of dogmas of economic or other fundamentalisms for which 'reality therapy' provided by Science may be useful. To be sure, the ruling elites have access to advice from scientists, but advice is no substitute for personal awareness or understanding of Science as the essential human activity.

The problem with economic fundamentalism (as with all other fundamentalisms) is not its pursuit of economic development as such. The economic development has considerably improved the lives of humanity! The problem is that economic fundamentalism does not recognize limits, boundary conditions and intrinsic interactions within any dynamical system like human society. In this respect fundamentalism can be considered an excellent example of reductionism; everything is reduced and subject to a single idea or a concept.

Reflection:

Does Science (in your opinion) have a future and why or why not?

Why do many people today have ambiguous (love-hate) attitude towards Science?

Tips: consider whether human society has a future or not. Consider the results of Science which have potential impact on Society.

Limits of Science [13]

Rapid scientific advances of the 20th century posed the question whether we are approaching the limits of scientific knowledge. Have we approached the 'Theory of Everything' and shall we soon become engaged only in fine tuning of our existing knowledge? We have stated earlier in this Chapter that at present Science still has a lot of scope for development and problem solving [11]. The limits of Science at present may have more to do with the limitations of scientists than with the availability of things that can be investigated. Nevertheless, in view of tremendous advances made in our understanding of Nature the question remains whether at some future time Science may come to an end because we shall have acquired the complete scientific description of natural world.

There are two sets of views regarding this problem.

- The first view is that a complete description is indeed possible and that all the fundamental questions will one day be answered. Thus, scientific activity in e.g. physics shall become a matter of technological development.

- The second view is that Nature is inexhaustible source of research problems because it continuously evolves and changes. Natural world is not a completely defined or developed entity. It continues to show diversity, so that the dialogue between human consciousness and the Universe will never end.

When we keep refining our descriptions, theories and models, do we reach the point when refinements become simply fine tuning? There is another aspect pertaining to the limits of Science which needs to be mentioned. There is recent awareness that some questions we pose about Nature exceed our technological ability to test them. Examples of such questions are incorporated in Big Bang Theory or Multiverse Theory. They also involve the structure of subatomic particles ('string theory') and physical events in the early Universe. Nevertheless, it is the triumph of modern physics that we can pose such fundamental questions in a rational, mathematical way. Let us consider fundamental limits on the development of Science which cannot, even in principle, be breached.

Human and social limits

We can suppose that human limits on Science exist because our minds were perfected in the evolutionary struggle for survival; our minds were not designed for the purpose of understanding the World. In other words, the development of our mental capabilities is an 'overkill' or by-product of Evolution. This suggestion may sound like a denigration of the value of human beings, but we need to remember that Evolution is not a process which has a specific purpose. To recap, Evolution is a process of selection and survival which operates on populations of organisms and is driven by constraints imposed by the Environment. The Environment is not a conscious, intelligent subject, but a set of boundary conditions which apply to a given dynamical system. Nevertheless, we have made great progress in our understanding of Nature by combining simple ideas like counting, cause-effect into generally valid laws.

We can amplify the capabilities of our brains by artificial means (computers and computer networks) and by collective activity of members of scientific community. Nevertheless, the problem of tractability remains. The time required to solve a problem by executing a sequence of very large number of small steps (even when we fully understand each individual step) may be too large to be feasible e.g. the task of factorization of large or prime numbers. Furthermore, the reliance on scientific networks and emphasis on collaborative work may reduce diversity/originality of scientific ideas. The need to alleviate consequences of unrestricted economic/technological development may also re-direct efforts of Science towards solving practical problems rather than towards curiosity driven investigations and expansion of fundamental knowledge. For example, biomedical research oriented towards solving health problems is today one of the most active and best funded research areas and so are the ecological studies. Democratic societies are reluctant to devote large portions of their GDP to projects which have no direct practical benefits. The projects involving CERN collider may be exceptions. This type of social priorities is reflected in questions which university students often ask their lecturers: 'What is the purpose of this or that particular exam or tutorial question?'

Technological limits

There are also technological limits on scientific development. They include:

- how fast can the acquired scientific information be transmitted and processed
- how accurately can we measure time and space (limits on nanotechnology)
- how much energy is required to gather scientific information
- how sensitive are technological devices to experimental errors and to chaotic amplification of uncertainties

Cosmological limits

Cosmological limits exist due to the finite speed of light which transmits information. This fact permits us to determine the structure of only that part of the Universe from which light can reach us ('visible Universe'). Because of the speed of light limit we cannot be sure that Universe is the same everywhere. We cannot tell whether Universe is finite or not, whether it had the beginning or end. Another example of this limit is the so called 'dark matter' which is difficult to study since its information carriers (gravitational fields or gravitons) are difficult to detect.

Logical/mathematical limits

There are also deep limits related to our mathematical description of the World. Mathematics cannot be proven to be both consistent and complete (remember *Gödel theorem,* Chapter 2). In our scientific investigations we are tied to the assumption that Nature is logically consistent which may leave certain aspects of Nature hidden from us. The assumption about logical consistency has served as well, but it is a product of inductive reasoning which (as discussed in Chapter 2) is not completely satisfactory.

Reflection:

Give an example of the question relevant to you which Science cannot answer and explain. Why Science cannot explain it? Do you think that Science shall be able to answer that question in the future? Give reasons for your answer.

Hint: consider for example the question of the meaning of Life

Science in Action-Case studies

To better understand a complex activity like Science it is best to observe it 'in action'. Three historical case studies which follow were selected to illustrate general characteristics of modern Science as an activity.

Case Study 1: Theories of combustion [14]

Development of Theory of combustion is the subject of science play, *Oxygen* [15]. Combustion is a very common and important process with which humans have been familiar since the Stone Age. Yet it took many millennia for us to understand it!

The theory

In the 17th century, phlogiston theory was proposed by G. Stahl which stated that all flammable objects contain the substance 'phlogiston' (from the Greek *phlogizo*, meaning 'to inflame'). According to this theory, combustion is the process which involves *losing material from the burning body*. This theory agreed with and explained many observations such as burning of coal, wood and plant materials. In all these cases there is a decrease in the amount of original substance present, leaving only a small amount of ash.

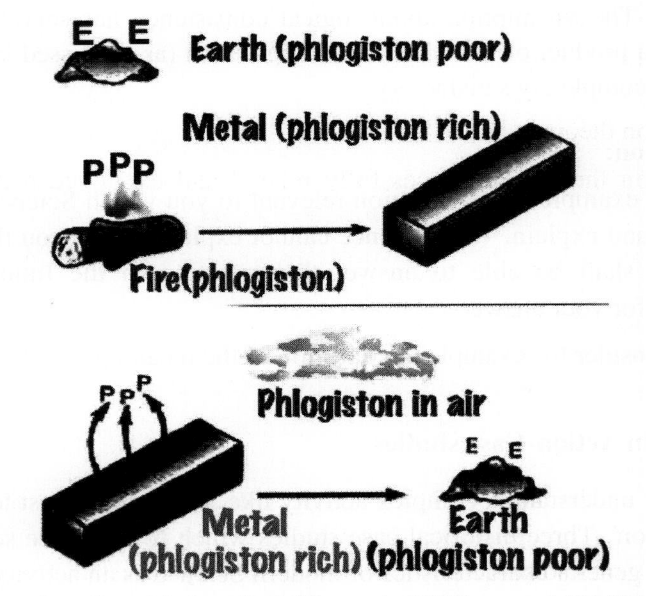

source: http://honolulu.hawaii.edu/distance/sci122/Programs/p24/p24.html

The pictorial diagram above represents basic ideas of phlogiston theory on the examples of metal ore smelting and rusting.

Left: earth (iron ore; phlogiston poor) combines with fire (pure phlogiston) and gives metal (phlogiston rich).

Right: metal (phlogiston rich) releases phlogiston into the air and leaves rust (earth: phlogiston poor).

Here are few processes described by 'phlogistonic equations':

Burning of sulphur:
sulphur = sulphuric acid +phlogiston

Rusting:
iron (metal) = iron calx (rust) + phlogiston

Metallurgy:
iron ore (calx of iron) + phlogiston = iron (metal)

Dissolution of metals in acids:
metal(calx & phlogiston) + acid = salt(calx & acid) + phlogiston

Rusting (according to modern understanding) is the reverse process of metallurgical extraction of metals. The same conclusion is deduced from phlogiston theory.

Phlogiston theory has successfully related and explained a number of observations by reference to natural laws only, without invoking supernatural speculations. It was the first scientific theory in Chemistry.

The inadequacies of phlogiston theory

In the 18th century, phlogiston theory had run into difficulties. It was known that air is involved in combustion e.g. when a jar is placed over a lighted candle, the candle becomes extinguished. This observation ran contrary to the notion of substance (phlogiston) leaving the burning body. However, phlogiston theory was not discarded immediately, because it was so successful in explaining many observations. Instead, an *auxiliary hypothesis* was formulated as follows:

The air absorbs phlogiston; when the air is saturated with it the burning stops, because no more phlogiston could be released from the burning body into the air.

Another difficulty with phlogiston theory was how to explain the fact that metals gain mass on rusting. Hence, another auxiliary hypothesis was introduced as follows:

Phlogiston has negative mass so, upon rusting, the phlogiston leaves the body making it heavier.

This hypothesis is a good example of 'conservativism' in Science (see Chapter 2) which tends to modify the existing paradigm rather than devise a new one even if it means introducing strange concepts like 'negative mass'. In the mid-18th century, three chemists (Cavendish, Priestley and Scheele) studied gases and air experimentally (pneumatic chemistry). They concluded that air is a mixture consisting of at least two components (airs) which in modern terminology would be called gases:

- **dephlogisticated air (oxygen)** – this 'air' was thought to have great affinity for phlogiston and strongly supports combustion.
- **inflammable air (hydrogen)** – Schelle considered this 'air' to be pure phlogiston because it is released upon dissolution of metals in acids.

Scheele observed that silver chloride (AgCl) turns into metallic silver when exposed to light. Silver is a metal but AgCl is not, because silver metal dissolves in strong acids while AgCl does not. Silver as a metal must then contain phlogiston while AgCl as a salt does not. So where does the phlogiston in Ag come from? From light? Is light phlogiston rich?

Silver chloride + Light (phlogiston) = silver (metal)

Does phlogiston represent two different things at the same time: light and hydrogen gas? Does light have negative mass? These questions reflected confusion in the state of scientific knowledge at the time and a new thinking was called for.

The paradigm shift

The modern theory of combustion took a long time to develop because at the time:

- the existence of atoms and molecules was not accepted/known
- the composition of air was unknown but it was thought to be a single substance
- the notions of acids and bases were unclear

A. Lavoisier (1743-1794) proposed the alternative chemical theory in 1777 on the basis of rigorous quantitative measurements. He weighed all substances before and after the reaction. His theory stated that:

- bodies burn only in oxygen (**'air vital'**);
- air is used in combustion and the gain in mass by the substance burnt equals the loss of mass shown by the air (conservation of mass);
- in the combustion process, the combustible substance is usually changed into acid while metal goes into calx (oxide) i.e. metal + oxygen (air) = **oxide**

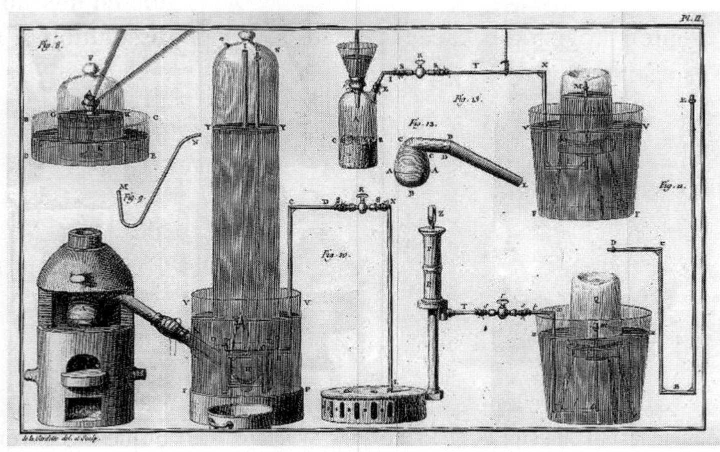

source: http://www.library.usyd.edu.au/libraries/rare/modernity/images/lavoisier2-1.jpg

Fig. 48.

Lavoisier performed his experiments using water trough to isolate gases which develop during chemical reactions (Figure 48). Lavoisier's theory was able to explain all the observed facts regarding oxidation and is still valid. He has also studied respiration, he designed calorimeter (for measuring energies in chemical reactions) and proposed logical nomenclature of chemical compounds based on their composition (e.g. oxides contain element oxygen). His combustion theory did not rely on the atomistic description of matter.

Lavoisier's book, *'Traite Elementaire de Chimie'*, published in 1789 was a milestone in modern Chemistry (Figure 49). Lavoisier participated in public life as a member of tax collecting company and was executed during French Revolution (see quotation below).

[On the execution of Lavoisier] Only a moment to cut off that head, and a hundred years may not give us another like it.
J.L. Lagrange

TRAITÉ
ÉLÉMENTAIRE
DE CHIMIE,
PRÉSENTÉ DANS UN ORDRE NOUVEAU
ET D'APRÈS LES DÉCOUVERTES MODERNES;
Avec Figures :
Par M. LAVOISIER, de l'Académie des Sciences, de la Société Royale de Médecine, des Sociétés d'Agriculture de Paris & d'Orléans, de la Société Royale de Londres, de l'Institut de Bologne, de la Société Helvétique de Basle, de celles de Philadelphie, Harlem, Manchester, Padoue, &c.

TOME PREMIER.

A PARIS,
Chez CUCHET, Libraire, rue & hôtel Serpente.

M. DCC LXXXIX.
Sous le Privilège de l'Académie des Sciences & de la Société Royale de Médecine.

source:
http://neptune.labunix.uqam.ca/CHI1515/Pages_Web/lavoisier21/contenu_files/image003.jpg / http://www.1902encyclopedia.com/C/CHE/traite-elem-lavoisier.jpg

Fig. 49.

What can we learn from Case Study 1?

Scientific theories evolve within the context of existing knowledge, but they also lead to new discoveries and ideas (see the following concept map).

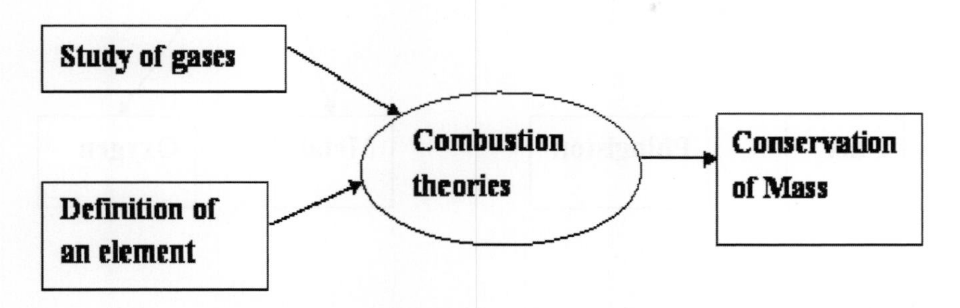

This case study is an example of a paradigm shift (see Chapter 2) which occurred in chemistry as shown schematically below.

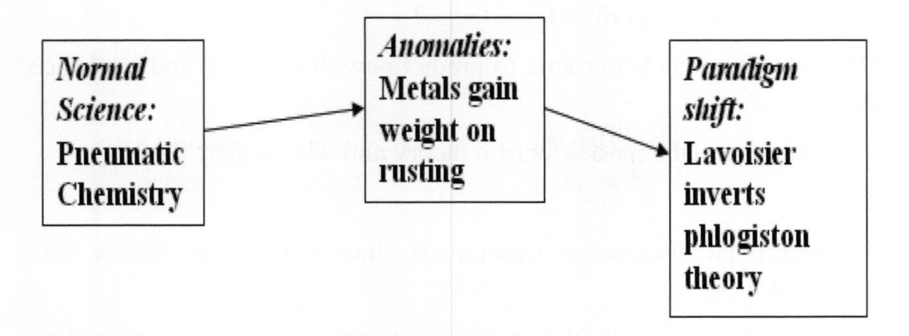

Consider the following two concept maps. The oxygen theory is preferred on the basis of *Ockham's razor* principle. This is because oxygen theory has greater explanatory coherence and explains most observations with less complexity than does phlogiston theory. For example, oxygen theory does not require auxiliary hypothesis like 'negative mass'.

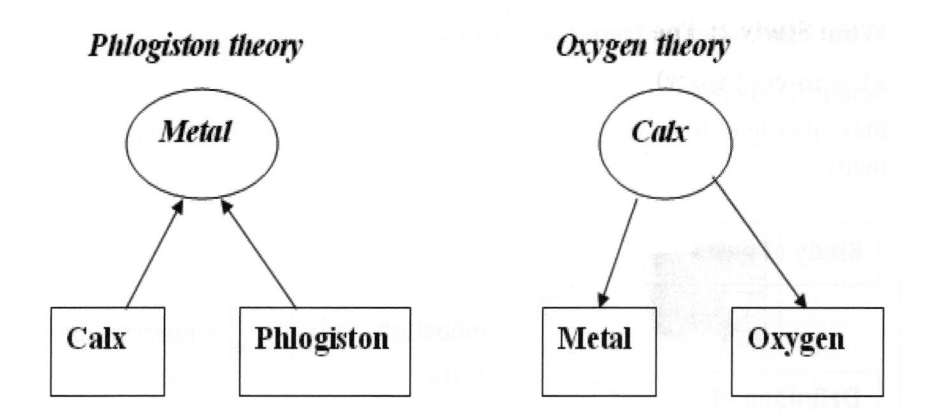

Reflection:

Did phlogiston and oxygen theories rely on the same pool of existing knowledge?

Which had a wider pool of knowledge?

Which theory was better able to predict new discoveries and introduce new ideas?

Was phlogiston theory a scientific theory and why or why not?

Tips:

consider what discoveries were made since phlogiston theory was proposed

consider what would be the prediction of phlogiston theory regarding human respiration

consider which scientific method was used in proposing phlogiston theory

Case Study 2: The Periodic Table of Elements [14] (Chemical jigsaw)

Periodic Table of the Elements

	IA	IIA	IIIB	IVB	VB	VIB	VIIB	VII	VII	VII	IB	IB	IIIA	IVA	VA	VIA	VIIA	0
1	1 H																	2 He
2	3 Li	4 Be											5 B	6 C	7 N	8 O	9 F	10 Ne
3	11 Na	12 Mg											13 Al	14 Si	15 P	16 S	17 Cl	18 Ar
4	19 K	20 Ca	21 Sc	22 Ti	23 V	24 Cr	25 Mn	26 Fe	27 Co	28 Ni	29 Cu	30 Zn	31 Ga	32 Ge	33 As	34 Se	35 Br	36 Kr
5	37 Rb	38 Sr	39 Y	40 Zr	41 Nb	42 Mo	43 Tc	44 Ru	45 Rh	46 Pd	47 Ag	48 Cd	49 In	50 Sn	51 Sb	52 Te	53 I	54 Xe
6	55 Cs	56 Ba	57 *La	72 Hf	73 Ta	74 W	75 Re	76 Os	77 Ir	78 Pt	79 Au	80 Hg	81 Tl	82 Pb	83 Bi	84 Po	85 At	86 Rn
7	87 Fr	88 Ra	89 +Ac	104 Rf	105 Ha	106 Sg	107 Ns	108 Hs	109 Mt	110 110	111 111	112 112	113 113					

* Lanthanide Series	58 Ce	59 Pr	60 Nd	61 Pm	62 Sm	63 Eu	64 Gd	65 Tb	66 Dy	67 Ho	68 Er	69 Tm	70 Yb	71 Lu
+ Actinide Series	90 Th	91 Pa	92 U	93 Np	94 Pu	95 Am	96 Cm	97 Bk	98 Cf	99 Es	100 Fm	101 Md	102 No	103 Lr

source: http://www.can-do.com/uci/lessons98/I-periodic-color.gif

After chemical elements were first identified as the smallest building blocks of matter in the 18th century, the similarities between properties of some elements was noticed. These included similarities in:

- **physical properties** of elements (aggregation state-metal, solid, liquid, gas, melting point temperature, boiling point, density, crystal structure)
- **chemical properties** of elements (valency, reactivity towards other elements)

For example, several elements (alkali metals) have high reactivity, chemical valency of one, low density, low melting points. Some elements form compounds with analogous (isomorphic) crystal structures e.g. NaCl, KCl, LiF, NaI. Furthermore, the discoveries of new elements in the 19th century raised the question of how many chemical elements there are altogether.

In the early 19th century, atomic masses were difficult to measure and hence sometimes inaccurate. Several scientists have proposed periodic relationships between chemical properties of elements and their atomic masses. The most important and far-reaching contribution was made by D.I. Mendeleev who published the first modern version of the Periodic Table in 1869.

source: http://library.thinkquest.org/05aug/01087/Mendeleev2.gif

His Table was not the first to be proposed, but Mendeleev's contemporaries could not explain discrepancies found in their tables. Mendeleev compiled all the available information about elements and by trial-and-error discerned the periodic pattern. But his contribution went much further! He was confident that his periodic table represented deep order/regularity in Nature, and could thus be used to make predictions. His predictions are summarized below.

- He claimed that the atomic masses of several elements were incorrectly determined because they did not fit into his system.
- He attributed 'blank spaces' in his table to new, yet undiscovered elements whose properties he predicted using interpolations from his Table.

Mendeleev's predictions of new elements and their properties were subsequently verified experimentally by discoveries of these new elements. His version of the Table was then accepted and he received most of the credit for the introduction of Periodic System in Chemistry.

The Periodic Table organizes and rationalizes huge amount of chemical information and ranks as one of the great unifying principles (theories) in Science. It is comparable to the Theory of Evolution and Newton laws of motion. The amazing thing about the Periodic Table is that it was set up before the existence of atoms was accepted or the inner structure of the atom understood.

What can we learn from Case Study 2?

Scientific theories must have organizing and predictive capability and not just the explanatory capability. Periodic Table predicted new information and was thus open to falsification if the experiments did not discover the predicted elements. This fits well with Popper's idea that good theory should predict something which we do not know so that it can be falsified (see Chapter 2). Also, this case study shows that scientific theories are bridges along which Science progresses; they provide coherence and integrate scientific knowledge and information (see concept map and citation below).

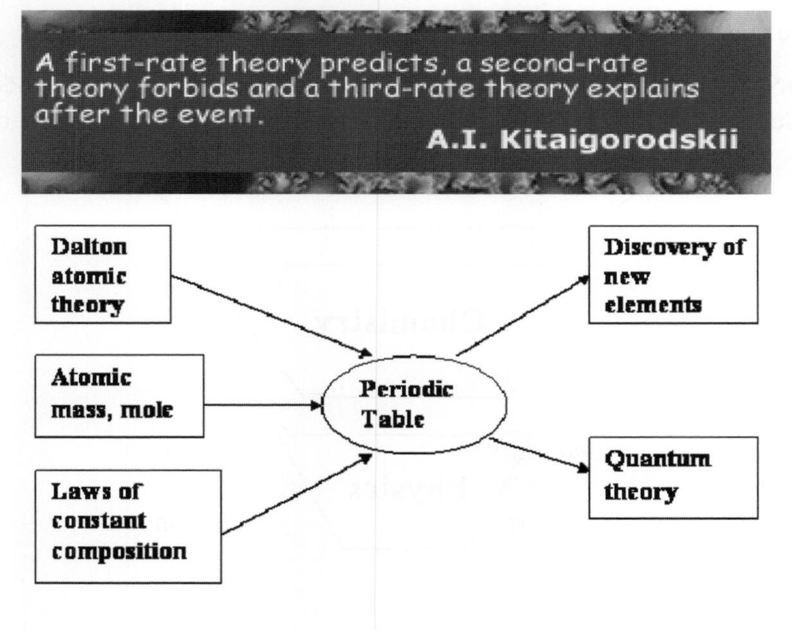

Scientific discoveries do not necessarily follow prescribed paths. Serendipity may also play a role as mentioned in Chapter 2! The organizing principle of the Periodic Table should actually have been based on atomic numbers, rather than on atomic masses. Serendipitously, the two properties follow each other closely i.e. the higher the atomic number the higher the mass. However, Mendeleev could not have known about atomic numbers because the structure of the atom was not discovered when he proposed the table. Yet he trusted the methodological principle that a coherent pattern which emerges from a vast amount of experimental evidence must be real, minor discrepancies notwithstanding.

The arrangement of elements in the Periodic Table was fully explained by Physics (Quantum Mechanics). This explanation exemplifies *reductionism* i.e. the description of more complex entity (atom) in terms of simpler entities (sub-atomic particles).

Can the rest of Science also be reduced to Physics?

(see concept map below)

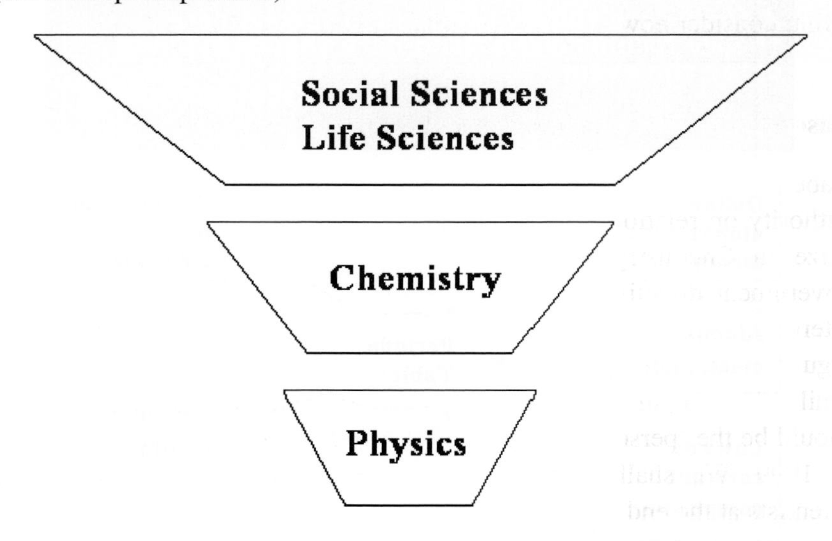

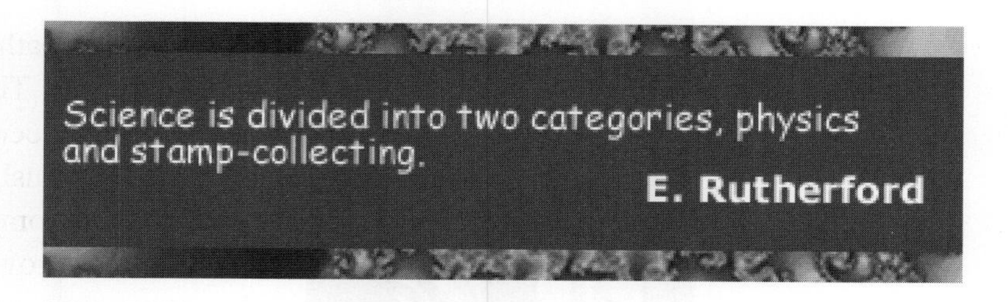

Science is divided into two categories, physics and stamp-collecting.
E. Rutherford

The above quotation by Rutherford is an amusing example of reductionist thinking. It claims that physics is the only fundamental science, that other sciences can be derived from it and are therefore mainly descriptive ('stamp collecting'). In a recent article Pross [16] has discussed another example of reductionism (biology to chemistry) which establishes a connection between laws of chemical kinetics and biological evolution, especially the emergence of Life. His article is an interesting attempt to discover 'chemical roots of Darwinism' [16].

Reflection:

Which crucial activity of Science (with respect to the processing of information) does Case Study 2 exemplify?

Hint: consider how are theories created and used in Science

Case Study 3: The life and work of Fritz Haber [17,18]

Haber was one of the leading physical chemists of his time and a world authority on relations between science and industry. He won the Nobel Prize in Chemistry and served his country unreservedly, accepting government directives without question. Let us take a look at how his interesting career unfolded. Haber and his compatriot Bosch were arguably 'the most influential people of 20th century' as suggested by Smil [17]. You may find this claim surprising and consider that Einstein should be the 'person of the century' according to Time magazine report in 1999. We shall compare the influences of these two prominent scientists at the end of this Chapter.

Photo source:
http://www.rsc.org/images/feature%20dronsfield%20fritz%20haber_tcm18-85279.jpg

Fritz Haber (1868-1934)

1906: Professor of Physical Chemistry, University of Karlsruhe

1911: Director, Kaiser Wilhelm Institute for Physical Chemistry, Berlin (KWI)

1918: Nobel Prize in Chemistry for the development of Haber-Bosch process (whereby ammonia is formed from hydrogen and atmospheric nitrogen under conditions of low temperature and high pressure).

1933: Resignation from his post as KWI Director. Emigration from Germany to Cambridge, UK

Haber was born and educated in Germany. In 1891 as a young undergraduate, he was presumably the first to synthesize a substance called 3,4-methylenedioxy-methamphetamine, or MDMA (known today as 'Ecstasy'). He gained his PhD in organic chemistry and held both industrial and academic positions. Chemical formulas of MDMA (Ecstasy), which Haber synthesized as a young undergraduate and his apparatus for laboratory synthesis of ammonia are shown in Figure 50.

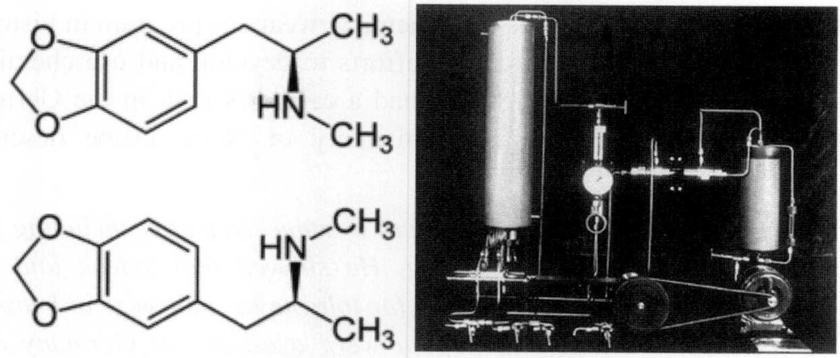

source: http://people.clarkson.edu/~ekatz/scientists/haber_apparatus.jpg

Fig. 50.

Haber later switched research interests from organic to physical chemistry and joined the Kaiser Wilhelm Institute for Physical Chemistry in 1911. He worked on projects in both pure and applied chemistry, including:

- fuel cells,
- glass electrodes
- electrochemistry
- thermodynamics

Haber's life's work was the ammonia synthesis which can be described by a simple chemical equation: $N_2 + 3H_2 = 2NH_3$. He also worked on applied chemistry problems and developed the first high pressure, catalytic industrial process.

The synthesis of ammonia from the elements nitrogen and hydrogen which are cheap and abundant was a problem whose solution had economic, political and military ramifications. Ammonia can be easily converted into nitrates or urea which is then used in the manufacture of huge quantities of fertilizers and explosives. It is interesting to note that the chemical mechanism of Haber's synthesis was only unraveled in 1977 by German chemist Ertl, many years after Haber's discovery of the process.

Haber was also 'father' of the first chemical weapons program in history. He supervised the German Army's efforts to develop and use chemical weapons in WWI (1914-1918) and had a captain's rank in the German Army. The lecture notes from University of Notre Dame describe his work as:

Haber's institute worked on numerous wartime concerns including the problem of keeping motors running. He showed that xylene and the solvent naptha were good substitutes for toluene as antifreeze in benzene motor fuel. Since xylene and naptha were available in Germany and toluene was not, Haber's contributions helped to keep German machinery running and aided in sustaining their war effort for four years.

Haber also served his country in the most basic sense with his process of ammonia synthesis. Not only was ammonia used as a raw material in the production of fertilizers, it was also (and still is) absolutely essential in the production of nitric acid. Nitric acid is a raw material for the production of chemical high explosives and other ammunition necessary for the war. Having helped to make Germany independent of Chile and other countries for necessary materials, Haber perhaps served his country in the greatest capacity. Without his process, Germany would never have had a chance to win the war.

Another contribution Haber made to Germany's war effort was in the development of chemical warfare. With strong purpose and great energy he became involved in the production of protective chemical devices for troops and directed the first gas attacks against enemy troops. Haber is often referred to as the father of modern chemical warfare as he organized and directed the first large scale release of chlorine gas on the western front at Ypres, France on April 22, 1915.

Source of notes: http://www.nd.edu/~nsl/Lectures/phys20061/pdf/2.pdf

This description shows how the ethical dilemmas of chemists preceded those of physicists who were involved in the development of nuclear weapons as part of the Manhattan Project. Haber had a simple solution to his ethical dilemma (see his statement which follows). It is hypocritical to pretend that similar sentiments do not exist today amongst leading elites and sections of populations in many countries today (scientists working for military-industrial complex)

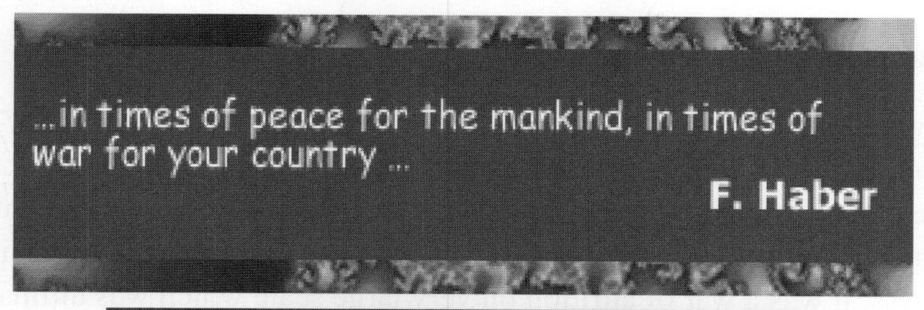

source: http://www.fu-berlin.de/tour/geschichtsausstellung/media/fritzhaber.jpg

Fig. 51.

The Kaiser Wilhelm Institute for Physical Chemistry in Berlin, where Haber worked, started in applied research, but from 1914 also performed weapons research (Figure 51).

Weapons of mass destruction

Why was there a need for chemical weapons? In World War I (WWI) there was a need for a new type of weapon which would break the stalemate of trench warfare. The warfare was causing enormous casualties without any significant gain of enemy territory or military success. It was a war of attrition on very large scale which was ultimately unsustainable. The belligerent nation which could quickly introduce a new weapon and break through enemy trenches had a good chance of military victory.

Haber's wife's (Figure 52) opinion on the matter was clear. Clara Immerwahr-Haber was a scientist in her own right, and became the first woman to obtain a PhD in chemistry in Germany. However, she believed that Science should be used for constructive purposes, not for making weapons of mass destruction. Haber tried to hide his work on poison nerve gas from her, but clearly wasn't successful. During an attack near Ypres where poison gas was used and around 5000 were killed and around 2000 blinded, she committed suicide.

source: http://www.uh.edu/engines/epi2287.htm

Fig. 52.

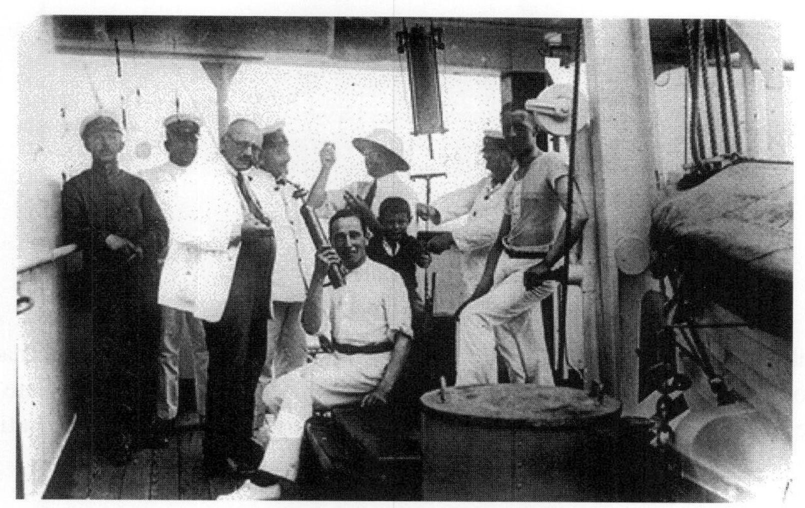

source: http://www.geocities.com/bioelectrochemistry/haber_group1.jpg

Fig. 53.

The design and use of weapons of mass destruction begun in WWI, continued in WWII (nuclear bomb) and the issues regarding their use are still unresolved (nuclear proliferation). Haber also worked, after WWI, on the unsuccessful attempt to extract gold from the sea to pay for German War Reparations (Figure 53; Haber aboard the research ship). The attempt was unsuccessful because unknown to Haber, the concentration of gold in the seawater is much smaller than he expected.

When Hitler came to power in Germany in 1933, the persecution of Jewish scientists was initiated as part of the racial program to 'purify' the German nation. Jews were portrayed as racially inferior and as socially harmful (as exploiting capitalists or as communist sympathizers). Haber the chemist, soldier and patriot became 'that Jew Haber'. He left Germany because there were constraints placed on his research and he was not allowed to choose his collaborators freely (see quotation which follows). He died in exile in spite of his great contributions to Science and to the Germany in WWI.

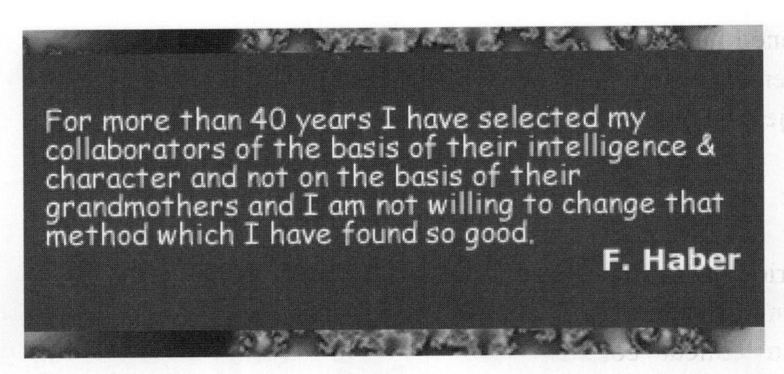

For more than 40 years I have selected my collaborators of the basis of their intelligence & character and not on the basis of their grandmothers and I am not willing to change that method which I have found so good.
F. Haber

Haber died in Switzerland in 1933. After 1933 great exodus of many prominent Jewish scientists to USA took place. The exodus was one of the reasons why after 1933, USA replaced Germany as the leading nation in scientific research.

The Two Scientists

What claim does Haber have to be the 'most influential person of 20th century'? Haber's design of the ammonia synthesis provided the source of cheap fertilizer and stimulated population growth in the World, especially in developing countries. Smil [17] calls Haber's work a 'detonator of the population explosion'. The extensive and complex consequences of this population growth we are all experiencing today. Great disparity in wealth between overpopulated and economically underdeveloped countries stoked up immigration pressures on the developed countries, leading to international conflicts, terrorism and degradation of the environment. The degradation is caused by increased need for agricultural land to feed the increasing population. This is a good example of unforeseen and unpredictable consequences which accompany many scientific discoveries. One must not conclude from this case study that scientific discoveries should be curtailed! Haber's work has provided food for many people who can, through their efforts greatly enrich modern civilization. The cause of these problems is not only the population growth (which does however have environmental limits), but also the political and economic circumstances in which this growth happened i.e. global distribution of wealth and power. One can add that another extremely influential chemical discovery, the contraceptive pill by Djerassi in 1951, skewed the World population distribution even further by significantly reducing the population growth in developed countries. 'Nature abhors gradients' and will sooner or later dissipate this gradient. How this will happen is an open question. The developed countries form the most scientifically and economically productive section of the World population [19]. N.B. This is a statement of fact and not an ethical comment on how the world population should be distributed.

Einstein's work did not even remotely exert comparable impact on the world societies. It is possible to argue that Einstein became famous for the wrong reasons. He did not become famous amongst the general public for his great scientific research, which most people do not understand. Einstein's name was associated with nuclear processes, technologies and weaponry even though he did not do much work in nuclear physics himself. The first nuclear transformation was affected by Rutherford and not by Einstein who proposed the equivalence of mass and energy. Nuclear technology had a highly visible, spectacular effects in a limited number of fields (primarily dropping of atomic bombs on Japan), but these effects are nowhere near as sustained and complex as those set in motion by Haber's work. We are not claiming that Haber was a 'better scientist' than Einstein (Haber was not), only that Haber's work turned out to be more influential of the two. Einstein's publicly visible personal humanity expressed through espousing political efforts towards freedom and democracy, his intellectual brilliance and his endearing image of absent-minded professor contributed to his celebrity status. Haber on the other hand supported the authoritarian, nationalistic regime in Germany not to mention its military forces. Finally, it is also possible that the Theory of Relativity captured the imagination of European public who was alienated from traditional ethical and social values after the horrors of WWI. Rejection and collapse of traditional values can often result in embracing relativism and doubt. However, we leave it to the reader to draw its own conclusions in this matter. Nevertheless, the discussion regarding Haber and Einstein demonstrates once again the profound relevance of Science for the modern World and the need to scrutinize scientific applications. Scrutinizing is not the same as curtailing or apportioning blame. The great challenge of the modern World is how to make Science free to acquire new knowledge and at the same time ensure that this knowledge benefits all of humanity.

What can we learn from Case Study 3?

The following concept map may be useful in considering this question

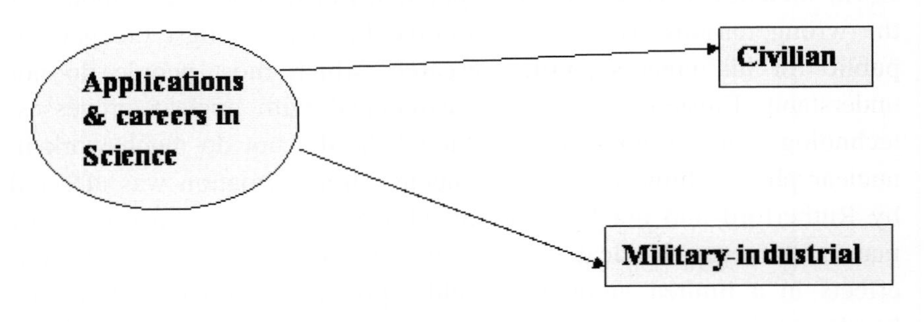

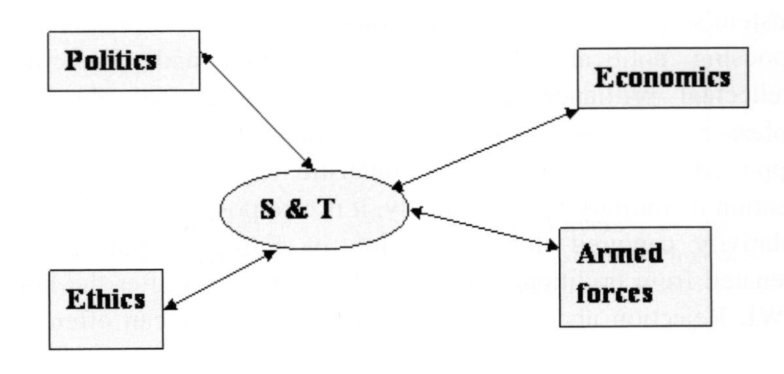

Case Study 3 illustrates several noteworthy points. The scientific activity coupled with technology lies at the heart of all developments in the modern World. Influence of ST is subtle and unpredictable, but is generally a lot more profound than more publicized artistic, political or economic events. Scientists are involved in all modern dilemmas and problems whether they like it or not. The consequences of scientific and technological advances cannot be predicted with certainty. Science is very pragmatic, it accepts theories, hypotheses and discoveries even if they are not fully understood as long as they can be used to further progress and new discoveries (c.f. the late elucidation of the mechanism of Haber process).

Reflection:

Science is often considered to be independent from social and political influences. What are the reasons why this is not possible in today's World? Does this imply that scientific results are swayed by social, political and economic developments?

Hints: consider the funding sources of modern Science. What are 'reference points' for scientific research and for other human activities?

Bibliography for Chapter 4:

1. a) Mackay, A.L. (1991) *A Dictionary of Scientific Quotations* (IOP Publishing, UK).
   b) Gaither, C.C., Cavazos-Gaither, A.E. (2000) *Scientifically Speaking* (IOP Publishing, UK).
   c) Alexander, D. (2001) *Rebuilding* the *Matrix* (Lion Publishing, UK).
2. McClellan, J.E., Dorn, H. (2006) *Science and Technology in World History*, 2nd ed. (The Johns Hopkins University Press, USA).
3. Dobson, G.P. (2005) *A Chaos of Delight, Science, Religion and Myth and the Shaping of Western Thought* (Equinox Publishing Ltd, UK).
4. Bowler, P.J., Morus, I.R. (2005) *Making Modern Science, A Historical Survey* (The University of Chicago Press, USA).

5. Cohen, H.F. (2007). Reconceptualising the scientific revolution, European Rev. 15, pp.491-502.
6. a) Diamond, J.D. (1998). Peeling the Chinese onion, Nature 391, pp.433-434. [29 Jan issue].
   b) Diamond, J.D. (1998) *Guns, Germs and Steel* (Vintage, UK).
   c) Diamond, J.D. (2005) *Collapse* (Penguin, UK).
7. Temple, R. (2002) *The Genius of China* (Prion Books Ltd, UK).
8. Sivin, N. (1982) Why the Scientific Revolution Did Not Take Place in China or Didn't It?, Chinese Sci., 5, pp.45-66.
9. Staal, F. (1995) Concepts of science in Europe and Asia, Interdiscip.Sci.Rev., 20, pp. 7-19.
10. Ernst, R.E. (2003). The societal obligations of Universities, Molecules, 8, pp.2-12.
11. Kennedy, D., Norman, C. (2005). What Don't We Know?, Science, 309, pp.75 [1 July issue].
12. Richards, J.R. (2000) *Human Nature after Darwin, A Philosophical Introduction* (Routledge, UK).
13. Barrow, J.D. *Impossibility. The Limits of Science and the Science of Limits.* (Vintage, UK).
14. a) Salzberg, H.W. (1991) *From the Caveman to Chemist, Circumstances and Achievements* (American Chemical Society, USA).
    b) Cobb, C, Goldwhite, H. (1995) *Creations of Fire, Chemistry's Lively History from Alchemy to the Atomic Age* (Plenum, USA).
    c) Levere, T.H. (2001) *Transforming Matter, A History of Chemistry from Alchemy to Buckyball* (The Johns Hopkins University Press, USA).
15. Djerassi, C., Hoffmann, R. (2001) *Oxygen* (Wiley-VCH, Germany).
16. Pross, A. (2009). Seeking the chemical roots of Darwinism. Bridging between Chemistry and Biology, Chem.Eur.J.,, 15, pp.8374-8381.
17. Smil, V. (1999). Detonator of the population explosion, Nature, 400, pp.415 [29 July issue].
18. Stoltzenberg, D. (2004) Fritz Haber, Chemist, Nobel Laureate, German, Jew (Chemical Heritage Press, USA).
19. Kirk, J.T.O. (2007) Science and Certainty (CSIRO, Australia).

Index